Profiled Sheet Roofing and Cladding

Profiled Sheet Roofing and Cladding
A guide to good practice

THIRD EDITION

Editor:
N.W. Selves BSc AIoR, Glanville Consultants

Editorial Committee:
J.J. Shanahan FIoR,
Formerly Chairman and Managing Director,
Wembley Roofing/Ceilings Ltd.
and
C.C. Sproul FIoR, Technical Officer,
National Federation of Roofing Contractors

London and New York

First published 1999 by E & FN Spon
11 New Fetter Lane, London EC4P 4EE

Simultaneously published in the USA and Canada
by Routledge
29 West 35th Street, New York, NY 10001

E & FN Spon is an imprint of the Taylor & Francis Group

© 1999 The National Federation of Roofing Contractors Publications Ltd.

Typeset in 10/12 Times by The Florence Group, Stoodleigh, Devon
Printed and bound in Great Britain by TJ International Ltd, Padstow, Cornwall

British Library Cataloguing in Publication Data
A catalogue record for this book is available from the British Library

Library of Congress Cataloguing in Publication Data
Profiled sheet roofing and cladding: a guide to good practice /
editor, N.W. Selves; editorial committee, J.J. Shanahan and C.C.
Sproul. — 3rd ed.
p. cm.
Rev. ed. of: Profiled sheet metal roofing and cladding. 2nd ed. 1991.
ISBN 0–419–23940–5 (pb)
1. Roofing, Iron and steel. 2. Sheet–metal, Corrugated.
3. Exterior walls. I. Selves, N.W. (Nicholas W.) II. Profiled
sheet metal roofing and cladding.
TH2457.P76 1999
695–dc21
98–45204
CIP

ISBN 0 – 419 – 23940 – 5

LIST OF CONTENTS

LIST OF CONTENTS   v

PREFACE   ix

INTRODUCTION TO THE 'BLUE BOOK'
– AN INDUSTRY OVERVIEW   xi

1. COMMON REQUIREMENTS FOR LIGHT
    WEIGHT PROFILED CLADDING   1

    1.1 Scope   1
         1.1.1 Products included in the guide   1
         1.1.2 Authority   1
    1.2 Performance of profiled sheeting   3
         1.2.1 General   3
         1.2.2 Dead load   3
         1.2.3 Imposed load   3
         1.2.4 Wind load   3
         1.2.5 Responsibility for load
                calculation   3
         1.2.6 Loads on lining sheets   3
    1.3 Performance of roofing and cladding   4
         1.3.1 Performance under load   4
         1.3.2 Openings in roofing/vertical
                cladding   4
         1.3.3 Purlin/side rail restraint   5
         1.3.4 Stressed skin performance   5
    1.4 Thermal movement   5
         1.4.1 Profiled sheeting   5

         1.4.2 Temperature range   5
         1.4.3 Accommodating thermal
                movement   6
    1.5 Deterioration – material incompatibility   7
         1.5.1 Material incompatibility   7
         1.5.2 Bimetallic corrosion   7
         1.5.3 Avoiding bimetallic contact   8
         1.5.4 Surface protection   8
         1.5.5 Latent deterioration   9
    1.6 Building Regulations – fire   9
         1.6.1 General   9
         1.6.2 Profiled sheeting and cladding   9
         1.6.3 CEN fire regulations   10
    1.7 Fire performance   11
         1.7.1 Fire requirements for walls   11
         1.7.2 Updating fire requirements   11
         1.7.3 Rooflights   11
         1.7.4 Loss Prevention Council   11
         1.7.5 FM approval   12
    1.8 Building Regulations – ventilation   12
         1.8.1 Approved Document F   12
         1.8.2 Internal environment   12
    1.9 Reducing the risk of condensation   12
         1.9.1 The role of the designer and
                contractor   12
         1.9.2 Air, moisture and condensation   13
         1.9.3 Movement of moisture through
                the construction   13
    1.10 Vapour control layers   14

1.10.1 The use of a VCL 14
1.10.2 Designers' responsibilities 14
1.10.3 Installing VCL 15
1.11 The breather membrane 15
1.11.1 Use of breather membrane 15
1.12 Thermal insulation 16
1.12.1 Choice of insulation 16
1.12.2 Design considerations 16
1.12.3 Thermal insulation – physical properties 16
1.12.4 Assessment of a system's thermal performance 17
1.12.5 Other insulated constructions 19
1.13 Air leakage 19
1.13.1 Approved Document L – Air Leakage 19
1.13.2 Air leakage testing 20
1.13.3 Method of test for air permeability of joints in building, BS 6181: 1981 20
1.13.4 BRESIM technique – BRE: Information Paper IP 11/90 20
1.13.5 References 20

2. PROFILED METAL ROOF AND WALL CLADDING 21
2.1 Introduction 21
2.1.1 Profiled sheet metal 21
2.1.2 Statutory and other requirements 21
2.1.3 Practical considerations 21
2.1.4 Durability 21
2.1.5 Appearance 21
2.1.6 Cost 22
2.1.7 Types of metal substrate 22
2.1.8 Steel substrate 22
2.1.9 Aluminium substrate 22
2.2 Durability 22
2.2.1 Life expectancy of profiled metal 22
2.2.2 Period to repaint decision 22
2.2.3 Life to first maintenance 22
2.2.4 Functional life 23
2.2.5 Selection by the specifier 23
2.3 Applications 23
2.3.1 Choice of profile 23
2.3.2 Use of light colours 23
2.4 Control of condensation 24
2.4.1 Hybrid construction 24
2.4.2 Internal humidity grades 24
2.5 Air leakage through the building envelope 24
2.5.1 Movement of air 24
2.6 Roof pitch and sheeting laps 25
2.6.1 Weathertight envelope 25
2.6.2 Side and end laps 26

2.7 Metal flashings 27
2.7.1 Design 27
2.7.2 Recommendations for the design of metal flashings 27
2.7.3 Roof penetrations 28

3. BUILT-UP METAL ROOFING SYSTEMS 30
3.1 Introduction 30
3.2 Assessment of the U-value for a cladding system 30
3.2.1 Built-up systems with spacers 30
3.3 Vapour control measures 33
3.3.1 Vapour control with insulated built-up metal sheeting 33
3.3.2 Installing VCL 33
3.3.3 Breather membrane 34
3.3.4 Hybrid construction 34
3.3.5 Vapour control and acoustic absorption 35
3.3.6 Vapour control and rooflights 35
3.4 Side and end laps 35
3.4.1 Profiled metal external sheeting 35
3.5 Secret-fix roofing systems 35
3.5.1 Definition 35
3.5.2 Thermal movement 36
3.5.3 Wind forces 36
3.5.4 Site practice and installing 36
3.6 Curved sheeting 37
3.6.1 Pre-formed curved sheeting 37
3.6.2 Reduced tolerance for fitting curves 37
3.6.3 Site work and appearance 38
3.6.4 Ordering details 39
3.7 Horizontally laid cladding 39
3.7.1 Sub-framing 39
3.7.2 Sheeting 40
3.7.3 Access 40
3.7.4 Maintenance 40
3.7.5 General 41

4. COMPOSITE PANELS 42
4.1 Introduction 42
4.2 Materials 42
4.2.1 The metal facings 42
4.2.2 The insulation core 43
4.3 Manufacturing 43
4.3.1 General 43
4.3.2 Continuous production 43
4.3.3 Batch production 43
4.3.4 Individual panel production 44
4.4 Panel performance 44
4.4.1 Structural 44
4.4.2 Thermal 44
4.4.3 Condensation 44

4.4.4 Acoustic 45
4.4.5 Fire 45
4.5 Structural 45
4.5.1 General 45
4.5.2 Design loading 45
4.5.3 Supporting structure 46
4.6 Roof applications 46
4.6.1 General 46
4.6.2 Traditional composite panels 47
4.6.3 Concealed fix composite panels 47
4.7 Standard details 48
4.7.1 General 48
4.7.2 Ridges 48
4.7.3 Eaves 48
4.7.4 Verges 48
4.7.5 Openings 48
4.8 Rooflights 49
4.9 Wall cladding 49
4.9.1 General 49
4.9.2 Roof systems as wall cladding 49
4.9.3 Finishes for wall panels 49
4.9.4 Purpose made wall panels 49
4.9.5 Supporting structure 50
4.9.6 Special details 50
4.10 Sitework and handling 51
4.10.1 Site storage 51
4.10.2 Site handling 51
4.10.3 Health and safety 51
4.10.4 Fixing requirements 51
4.10.5 Repairs to damaged panels 51
4.10.6 Inspection and maintenance 52

5. FIBRE-CEMENT SHEETING 53

5.1 Introduction to fibre-cement 53
5.1.1 Appearance 53
5.1.2 Properties 54
5.2 British Standards for fibre-cement products 54
5.2.1 Standards for profiled sheeting 54
5.2.2 Fittings 55
5.3 Recommendations for installation 55
5.3.1 Movement joints 55
5.3.2 Lap treatments 55
5.3.3 Design guidance procedure 55
5.4 Installation 57
5.4.1 General 57
5.4.2 Fixing 57
5.4.3 Weather protection 58
5.4.4 Overhangs 61
5.4.5 Side laps 61
5.4.6 End laps 61
5.5 Fixing procedure for laying with mitred corners 61
5.6 Notes for guidance on storage 62

5.6.1 Natural grey sheets 62
5.6.2 Coloured fibre-cement and metal lining sheets 63
5.7 Safety 63
5.7.1 Preparatory safety 63
5.7.2 Safety at work 63
5.8 Roofing systems 63
5.8.1 Metal lining system 64
5.8.2 Rigid installation boards 64
5.8.3 Double skin system 64
5.9 Materials and performance 64
5.9.1 Material characteristics 64
5.9.2 Chemical resistance 64
5.9.3 Thermal movement 64
5.9.4 Sound insulation 64
5.9.5 Condensation 65
5.9.6 Effect of frost 65

6. ROOFLIGHTS 66

6.1 Introduction to rooflights 66
6.2 Design aspects 66
6.2.1 Material and types 66
6.2.2 Daylighting levels 67
6.2.3 Weathering and durability 67
6.2.4 Typical application 67
6.2.5 Building Regulations 68
6.2.6 Loading – general 69
6.2.7 Safety and CDM 70
6.3 Site aspects 71
6.3.1 Fasteners 71
6.3.2 Sealants 73
6.3.3 Thermal expansion 74
6.3.4 Fixing sequence 74
6.4 Transport, handling and storage 74
6.5 Maintenance 74

7. FASTENERS AND FIXINGS 75

7.1 Introduction to fasteners and fixings 75
7.1.1 Primary fasteners 75
7.1.2 Secondary fasteners 75
7.2 Modes of failure 75
7.2.1 Principal modes of failure of fasteners in service 75
7.2.2 Standards and test methods 76
7.3 Types of fasteners 76
7.3.1 Primary fasteners 76
7.3.2 Secondary fasteners 77
7.3.3 Other fixing applications 78
7.4 Fastener materials 78
7.4.1 Choice of materials 78
7.4.2 Fastener heads 80
7.4.3 Exposed fasteners – colour matching 80
7.4.4 Fastener washers 80

7.5   Fastener application   81
      7.5.1   Forces acting on fasteners   81
      7.5.2   Designer's responsibilities   81
      7.5.3   Sheeting contractor's
              responsibilities   81

8. SEALANTS AND PROFILE FILLERS   83

8.1   Introduction   83
      8.1.1   Reasons for use of sealants   83
      8.1.2   Profile fillers   83
8.2   Functions and desirable properties
      of sealants   83
      8.2.1   Functions of sealants   83
      8.2.2   Properties required of sealants   84
8.3   Forms of sealant available   86
8.4   Application of sealants   86
8.5   Profile closures   86
      8.5.1   Profile fillers   86
      8.5.2   Vented fillers   87
8.6   Eaves closure foam fillers   87
8.7   Fire resistant closures   87

9. RAINWATER GOODS   88

9.1   Introduction   88
9.2   Design considerations   88
      9.2.1   BS 6367: 1983 Code of Practice
              for drainage for roofs and paved
              areas   88
      9.2.2   Design rates of rainfall   88
      9.2.3   Wind   89
      9.2.4   Snowfall   89
      9.2.5   Run-off   89
      9.2.6   General principles of design   89
      9.2.7   Calculation of flows in gutters   90
      9.2.8   Design of outlets   91
      9.2.9   Syphonic outlet systems   91
      9.2.10 Gutter fixing   92
9.3   Valley, parapet and boundary wall
      gutters   92
      9.3.1   Gutter shapes   92
9.4   External gutters   93
      9.4.1   Eaves gutters   93
9.5   Gutter joints   93
      9.5.1   Methods of joining   93
9.6   Gutter insulation   94
      9.6.1   Thermal requirements   94
9.7   Materials for gutters and rainwater
      pipes   94
      9.7.1   General   94
      9.7.2   Steel   94
      9.7.3   Aluminium   94
      9.7.4   Coatings for mild steel and
              aluminium   95
      9.7.5   Stainless   95
      9.7.6   Plastisol coated steel   95
      9.7.7   Glass fibre reinforced plastic   95

      9.7.8   Composite gutters   95
      9.7.9   Unplasticized polyvinyl chloride   95
      9.7.10 Fibre-cement   95
      9.7.11 Asbestos cement   95
      9.7.12 Cast iron   95
      9.7.13 Outlet protection   95
9.8   Installation   96
      9.8.1   Insulated gutters   96
      9.8.2   Training   96
      9.8.3   Final checking   96
      9.8.4   Workmanship   96
9.9   Staining   96
9.10 Leaks   96
9.11 Maintenance   96

10. CONSTRUCTION, SAFETY AND
     MAINTENANCE   97

10.1 Introduction   97
10.2 Site safety   97
      10.2.1 The Health and Safety at
              Work, etc., Act 1974   97
      10.2.2 Employer duties   98
      10.2.3 Employees' duties   98
      10.2.4 Safe systems of work   98
      10.2.5 Existing roofs   98
      10.2.6 Access   98
      10.2.7 Plant and scaffold registers   98
      10.2.8 Site safety meetings   98
      10.2.9 Welfare   99
10.3 Health and safety section   99
      10.3.1 Health and Safety Executive
              publications   99
      10.3.2 The Health and Safety at Work,
              etc., Act 1974   99
      10.3.3 The Management of Health and
              Safety at Work Regulations 1992   99
      10.3.4 The Construction (Health, Safety
              and Welfare) Regulations 1996   99
      10.3.5 The Construction (Design and
              Management) Regulations 1994   99
10.4 Transport, handling and storage   100
      10.4.1 Transport   100
      10.4.2 Handling   101
      10.4.3 Storage   101
10.5 Preparation   102
      10.5.1 Commencement of work on site   102
      10.5.2 Additional site hints   102
10.6 Maintenance   102
      10.6.1 Maintenance manual   102
      10.6.2 Maintenance inspections   102
10.7 Fault finding analysis   103

APPENDICES   105

Appendix A: British and European Standards   105
Appendix B: Reference Documents   107

INDEX   109

PREFACE
National Federation of Roofing Contractors: service to members and clients alike

The National Federation of Roofing Contractors (NFRC) intends this guide to be a description of current good practice in the application, design and installation of profiled sheets which are used in roof and wall cladding. This is the third edition of the NFRC's guide which aims to assist in increasing cooperation between the designer and contractor and to inform all members of the building team about the abilities and applications of products.

The third edition, like its forerunners, has been revised, drafted and re-drafted, with the enthusiastic cooperation of NFRC contractor and manufacturer members who contribute to the work of the Joint NFRC/ FRA Sheeting and Cladding Committee. The Committee has been under the chairmanship of Mr J.J. Shanahan throughout this period of revision.

It is intended that the guide should be used in conjunction with relevant British Standards (BS), European Standards (BS EN), and Codes of Practice which will supplement the guide on matters which relate specifically to profiled coated steel, aluminium, translucent rooflights or fibre-cement cladding and roofing. It must be clearly understood that this guide represents a description of what is generally considered to be good and prudent practice. Compliance with the guide does not confer immunity from statutory requirements, regulations or bye-laws, nor does it purport to describe the maximum tolerance or performance of products and materials. The guide is a reflection of an industry view of what constitutes sound practice for each of several general product types, by describing techniques based on experience and how construction works in practice.

The final responsibility for the whole design specification and for any judgement about the practicability of designs in the prevailing conditions must be with the designer. The satisfactory execution of the installation is the responsibility of the contractor. No recommendation or description of practice in this guide is intended to inhibit, or impinge upon the judgement of the designer, or to prevent a qualified designer or engineer from introducing new methods, and/or techniques, based on their individual skills and experience.

Trade members of the NFRC undertake both roof and wall cladding with profiled metal and fibre-cement sheeting. In this Guide the term 'Roofing Contractor' has been used to mean an experienced contractor who is a specialist in both roof and wall cladding with these materials.

WHAT IS THE NFRC ?

The National Federation of Roofing Contractors (NFRC) is the largest roofing trade federation in the UK which has contractor members who carry out work in all roofing disciplines including:

- industrial and agricultural sheeting and cladding;
- slating and tiling (including shingles);
- built-up felt and single ply roofing;
- mastic asphalt and liquid waterproofing systems;
- aluminium/copper/lead/steel/zinc roofing and decking.

There are more than 100 manufacturer and service provider NFRC associate members who can supply a comprehensive range of products, systems and support services to the client and his designers, engineers and contractors.

The NFRC exists to promote the interests of all its members, to provide skilled workmanship and quality products to the client and thus enhance the reputation of the British roofing industry. Associate members of the NFRC include the principal profiled cladding manufacturers, i.e. sheet metal roll-formers, translucent sheet and fibre-cement producers and ancillary product manufacturers. Their effort in industrial roofing and cladding depends on the close liaison, which they encourage between their contractor and associate members, to offer clients, developers and designers, commercial benefits and unique guarantees. For agricultural buildings, the close relationship between NFRC roofing contractors and sheet roll-formers or fibre-cement producer offers farmers, estate managers, and other clients similar benefits and guarantees.

The NFRC has invested its resources in professional research into the essential needs of the roofing industry to improve the standards of recruitment, initial and further training. Improved skill levels will increase cost benefits. They are also a necessary requirement to ensure that existing Health and Safety Standards are enhanced and that roofing contractors are capable of taking advantage of new product developments, techniques and systems.

This new edition of *Profiled Sheet Roofing and Cladding* is an important service to members, specifiers and clients alike, as is the Federation's second edition of its electronic disk publication, *The NFRC Guide to Quality Roofing*, which lists all trade and associate members who install or produce sheeting and cladding materials. The NFRC Internet web-site, www.nfrc.co.uk, carries the same listings.

The NFRC provides the unique Co-partnership Insurance Guarantee Scheme which covers a full ten year period and provides for:

- a guarantee from the NFRC member covering his work;
- a material guarantee from the NFRC associate manufacturer or merchant;
- an insolvency guarantee managed by the Insurance Guarantee Association (IGA), backed by underwriters to overcome the eventuality of the trade or associate member ceasing to trade.

Each contract is dealt with separately. The client receives a certificate of insurance and a written guarantee on completion of the work – these can be passed on to new owners and remain in force for the full ten year period.

The NFRC is developing further insurance backed guarantee schemes, including cover for latent defects.

ACKNOWLEDGEMENTS

The Editors and NFRC Publications Ltd gratefully acknowledge the assistance and support of the following in the preparation of this third edition:

The Metal Cladding and Roofing Manufacturers Association (MCRMA);

The Fibre Cement Manufacturers Association (FCMA);

The Association of Roof Light Manufacturers (ARM);

The National NFRC/FRA Joint Sheeting and Cladding Committee;

and many other members of the NFRC who have contributed advice and information.

Extracts from BS 5427: Part 1: 1996 are reproduced with the permission of BSI. Complete editions of the standards can be obtained by post from BSI Customer Services, 389 Chiswick High Road, London, W4 4AL.

Extracts from BS 5534: Part 1: 1997 are reproduced with permission of BSI under licence no. PD\1998 1678. Complete editions of the standards can be obtained by post from BSI Customer Services, 389 Chiswick High Road, London, W4 4AL.

Extracts and drawings from MCRMA's Technical Papers are reproduced with the permission of the MCRMA. Complete copies of Technical Papers can be purchased from the Metal Cladding and Roofing Manufacturers Association Ltd, 18 Mere Farm Road, Noctorum, Birkenhead, Merseyside, L43 9TT.

INTRODUCTION TO THE 'BLUE BOOK' – AN INDUSTRY OVERVIEW

There has been a tremendous advance in the sheeting and cladding industry since the NFRC last published its 'Blue Book' – *Profiled Sheet Metal Roofing and Cladding – A Guide to Good Practice* back in 1991. Innovation both in technology and in practice have moved at a pace. The result is that while the basic principles of good roofing remain unchanged, the variety of new products and systems that have been launched on to the market ensure that designers now have a wide range of alternative choices to call upon for any given project.

This new and improved version of the 'Blue Book' picks up on these various options and sets them in the context of industry good practice. This should ensure that all those interested in the industry are able to make the most of the systems available.

A CHANGING INDUSTRY

The bulk of this publication will look in detail at each of the key elements of profiled sheeting and cladding. But, in this short introduction, we will touch on the many positive changes that have affected the industry and that are covered in depth in each relevant chapter. They can be summarized as follows:

- better materials, sealants and fixings;
- increasing levels of insulation and a better understanding of the concept and effects of heat loss and condensation;
- growth in the use of composite panels;
- increase in the use of 'secret-fix' systems;
- the Construction (Design & Management) Regulations 1994 (CDM) – ensure designers have to be responsible for considering site safety;
- increased authority and focus from the Health and Safety Executive on all aspects of roofing safety – especially fragile materials and open-edge protection;
- greater response by industry to safety concerns – exemplified by increasing use of 'walk-on' liner panels and 'walk-on' rooflights as safe working panels;
- action to improve quality and durability of metal-clad roofs through initiatives including ZELRO, the Roofing Industry Alliance and Latent Defect Insurance Schemes.

While the first chapter of this publication covers subjects of general interest to designers, manufacturers and industry practitioners, the second chapter describes the common performance details when profiled metal is the weathering surface and covers durability, pitch and laps. The subsequent chapters focus in on more specific detail. This overview offers a 'short-hand' introduction to enable you to select areas of interest and relevance to your personal needs.

BUILT-UP SYSTEMS

The roof is typically 'built-up' by laying liner panels over purlins and then introducing a spacer grid system to support the weather sheet over loose laid insulation.

The benefits of such a system include:

- flexibility and ease-of-application to roofs of any shape;
- speed of 'lining-out' – enabling work to continue unhampered beneath;
- economy;
- favourable tolerance of structural steel movement or mis-alignment;
- the capability to be installed in long lengths utilizing a variety of 'secret-fix' arrangements;
- low cost local damage repair;
- a variety of profiles and suppliers;
- acoustic performance;
- non-combustible insulation.

Care points to consider include:

- condensation risks – liner panels must be sealed;
- management of components – all systems use a number of flashings, fixtures and components which require effective site control and good practice in their use;
- safety – especially if liner panels cannot be used as a working platform.

Chapter 3 covers built-up systems in detail.

COMPOSITE PANEL SYSTEMS

Since the last 'Blue Book' was published, composite panel systems have made significant advances and have gained in popularity. It is believed that the area being clad per year with composite panels has almost trebled since the last 'Blue Book'. Since the outer skin is not designed to be affected by conditions within the building, these roofs are less prone to condensation problems and are considered 'warm'.

The benefits include:

- simpler application since top sheet, insulation and liner sheet form one component;
- integrity of insulation;
- quicker access – as the roof can be walked on as soon as it is fixed to purlins.

Care points to consider include:

- project timings – work below can proceed only under completed sections of the roof;

- steel work – this must be installed to finer, closer tolerances due to the rigidity of panels;
- extra care necessary in transport, handling and installation phases.

You can read more about composite systems in Chapter 4.

FIBRE CEMENT

Fibre-cement sheeting offers a number of advantages – particularly in durability and condensation absorbency – which makes it an attractive proposition in many agricultural and industrial applications. To find out when it is most appropriate, read Chapter 5.

ROOFLIGHTS

There have been a number of recent developments in rooflights which are covered in detail in Chapter 6. In essence, the two key advances that benefit specifiers are:

- the availability of vaulted rooflight systems to complement 'secret-fix' and composite roofing systems;
- 'safe' rooflights, overcoming the historic risks (strength, weathertightness and durability) previously associated with translucents.

FASTENERS AND FIXINGS

Chapter 7 covers the detail of fixtures and fixings in 1998. It remains a key area which is not regarded as seriously as it might be. Many failures in profiled sheeting arise from the incorrect selection or wrong application of fixings. The NFRC supports the view that specifiers must bring the same clarity to their specification of fixings as they do to the cladding itself.

SEALANTS AND FILLERS

In Chapter 8, guidance is given on the choice of sealants and fillers used to close off cavities. It is clear that inadequate sealing of liner panels is the most important single factor in subsequent condensation problems. Good sealing will eliminate most problems and may well avoid any need for additional sealing membranes in built-up systems.

GUTTERS

In Chapter 9, Rainwater Goods, there are details on gutter materials, joints and insulation together with design criteria.

SITE PRACTICE

The greatest theory will count for nought if good site practice is not followed, and Chapter 10 will be essential reading for all involved in this area. While legislation is currently under review, the NFRC continues to advocate strongly for all specifiers and contractors to consider more carefully all aspects of safety, when designing roofs. This consideration should lead them to set out a programme to implement all aspects of safety. We fully expect the Health and Safety Executive to be more active over the coming years in enforcing safety issues.

MAINTENANCE

All the chapters provide useful guidance on roof maintenance. A roof can be a hostile environment, subject to UV radiation, wind, weathering and the effects of pollution. Clearly good maintenance is essential. Given the right specification and fitting – and the right level of on-going maintenance – a modern, metal-clad roof is capable of trouble free service for at least 30 years.

All roofs require at least one annual inspection to be scheduled. Maintenance service and repairs by professional roofers should be the response to any problems discovered.

The roofing industry is a dynamic environment. This edition of the 'Blue Book' sets out the changes that have occurred in the development of good practice up to the date of this publication. To follow on, the NFRC will be issuing Technical Bulletins and guidance to coincide with key changes as they occur and gain acceptance.

Information on price and availability of current publications is available from NFRC Publications Ltd, at 24 Weymouth Street, London W1N 4LX, Tel: 0171 436 0387 and Fax: 0171 637 5215.

Edward Cowan, OBE
Chief Executive, NFRC

1

COMMON REQUIREMENTS FOR LIGHT WEIGHT PROFILED CLADDING

1.1 SCOPE

1.1.1 Products included in the guide

The scope of this guidance document includes the following types of self supporting roof and wall cladding.

1. Single skin profiled metal sheeting of aluminium or steel (Figure 1.1).
2. Insulated profiled metal sheeting of aluminium or steel in site assembled layers (Figure 1.2).
3. Factory insulated composite panels [sandwich panels in accordance with European Committee for Standardisation (CEN) terms] with profiled metal skins (Figure 1.3).
4. Single skin profiled and site assembled insulated fibre-cement cladding (Figure 1.4).
5. Profiled rooflights intended for use as part of the above mentioned cladding systems.
6. Fitments and fasteners associated with these types of light weight cladding systems.

1.1.2 Authority

Reference is made in this Guide to Government Regulations, Standards and Advice Documents. The majority of those relevant to the roofing and cladding industry are listed in Appendix A with reference, if any, to the appropriate sections of the Guide.

The documents are listed by the following classifications.

- British Standard Documents
    British Standards BS
    European Standards BS EN
    Codes of Practice (CoP) CP
- Building Research Establishment (BRE) Documents
    BRE Digest
    BRE Reports
- Government Regulations
    Acts of Parliament

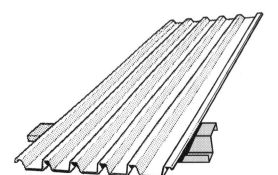

Figure 1.1 Single skin profiled metal sheeting of aluminium or steel (reproduced with permission from the MCRMA).

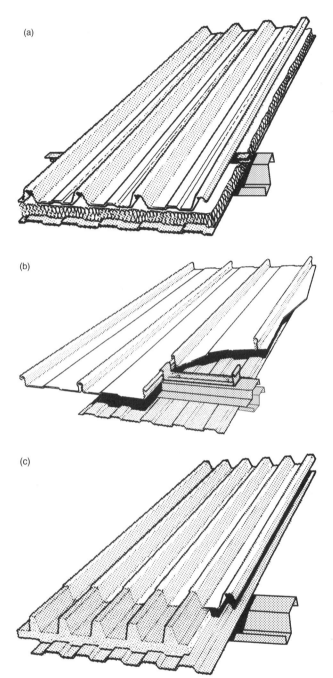

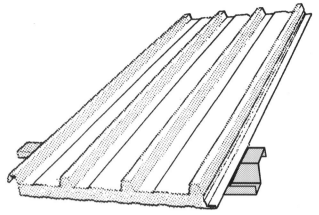

Figure 1.3 Factory insulated composite panels [sandwich panels in accordance with European Committee for Standardisation (CEN) terms] (reproduced with permission from the MCRMA).

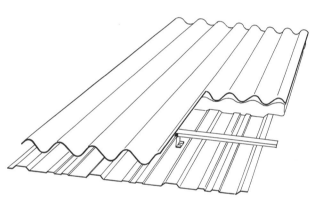

Figure 1.4 Single skin profiled and site assembled insulated fibre-cement cladding (reproduced with permission from FCMA).

Figure 1.2 Insulated profiled metal sheeting of aluminium or steel in site assembled layers: (a) built up insulated or hybrid construction; (b) secret-fix; (c) site assembled composite roof (reproduced with permission from the MCRMA).

The Member States in 1998 are:

Austria	Belgium	Czech Republic
Denmark	Finland	France
Germany	Greece	Iceland
Ireland	Italy	Luxembourg
Netherlands	Norway	Portugal
Spain	Sweden	Switzerland
	United Kingdom	

The traditional British Standards (BS) are being replaced by European Standards issued by CEN, the European Committee for Standardisation. Eventually, these Standards will be adopted and implemented in all member states of the European Union to allow free movement of products without the requirement for additional national testing or certification.

European Standards (BS EN) are issued by the British Standards Institute (BSI) and any corresponding British Standard is withdrawn. The BS ENs issued to date for the building industry are product standards. Other standards, similar to Codes of Practice, are expected to follow. Standards which are relevant, but still to be finalized, are listed as prEN.

1.2 PERFORMANCE OF PROFILED SHEETING

1.2.1 General

Part A of Schedule 1 to the Building Regulations 1985 requires buildings to be so constructed that all dead, imposed and wind loads are sustained and safely transmitted to the ground.

1.2.2 Dead load

Dead load is the permanent load due to the weight of the construction, its finishes and attachments.

In relation to profiled sheet roofing this will include the self-weight of the finished sheeting as a uniformly distributed dead load and any permanent attachment (e.g. roof ventilator) supported by the sheeting.

1.2.3 Imposed load

Imposed load, in the roofing context, is the temporary and variable assumed load (both uniformly distributed and concentrated) supported by the sheeting, but excluding wind load.

The British Standard Code of Practice for imposed loads is BS 6399: Part 3.

It will include snow load as a distributed load but with provision as appropriate for the effect of drifting snow in valleys and against vertical obstructions. In addition, the imposed load will include; crawl-boards for occasional roof access and concentrated loads attributed to operatives carrying out maintenance, repairs or alterations.

Dead and imposed loads are of little consequence in relation to vertical cladding, but consideration should be given to any horizontal force acting against parapet cladding, such as the result of sliding snow on an adjacent roof slope.

Cladding on a non-vertical wall should be treated as roofing but with due regard to the influence of its inclination.

1.2.4 Wind load

Wind load, in relation to roofing and cladding, must be assessed for each complete structural element (roof and wall) and for each sheet or panel and its fixings.

The British Standard Code of Practice for wind loads is BS 6399: Part 2. At the time of writing the code CP3: Chapter V: Part 2: 1973 may still be used, although it has been withdrawn by British Standards but not declared obsolete. A new European Standard is being prepared which may supersede both British codes.

The pressures exerted upon a building as the result of wind speed are influenced by many factors such as the height, proportions and orientation of the building, roof pitch and number of spans, nature of the general terrain, windbreaks and obstructions and effects of adjacent hills and valleys.

Air flow over and around buildings produces negative pressures (or suction) on much of the external cladding and roofing, but the total wind force on an element of the structure depends on the difference between external and internal pressures.

Openings in walls on the windward side of a building lead to an increase in internal pressure thereby adding to the effect of external suction and subjecting cladding/roofing fasteners to tensile loads and fixings to potential pull-out, pull-over and washer inversion forces. The use of light weight profiled sheet cladding provides relatively little by way of dead load to counteract this net outward wind created force.

Conversely, openings in areas subjected to external suction cause a reduction in internal pressure, adding to the total load on an element already under external pressure.

Attention should be given to the load on any roof overhang. The pressure acting on the underside will be related to the pressure on the adjacent wall surface, increasing the total load on the overhang, if that part of the roof is already subjected to external suction.

Although it is permissible to use less than the maximum design wind speed when considering the stability of partially clad structures during the temporary construction period, it may be found advisable, for buildability, to plan the sequence and method of working to minimize the risk of wind damage during construction.

1.2.5 Responsibility for load calculation

The designer must carry the responsibility for considering all modes of loading relevant to each area of cladding and roofing and to determine the performance requirements before completing their specification.

1.2.6 Loads on lining sheets

Stability of the linings is of great importance particularly when it forms a vapour control layer as the means of protection against interstitial condensation. Consideration should therefore be given to the loads imposed on internal lining systems whether using lining boards, metal liner panels or trays.

Account should be taken of the self weight of the lining material, the distributed dead load due to insulation (thermal and acoustic), other attachments and variations in internal pressure induced by the wind effects referred to above.

It is relevant to point out that a breather membrane over air-permeable insulating material, may be subjected to and possibly damaged by, air pressure acting outwards.

1.3 PERFORMANCE OF ROOFING AND CLADDING

1.3.1 Performance under load

A structural steel section used as a beam, derives its strength from the ability of its flanges to resist tensile and compressive forces, but this strength can only be maintained whilst those parts of the beam in compression are restrained from buckling.

A profiled sheet is in effect a complex beam in which the crown and valley material acts as the flanges with the sloping sides of the profile serving as the web, keeping the 'flanges' apart by the correct distance and allowing them to give the complete sheet its rigidity.

When a roof sheet deflects downwards under combined dead and imposed loads, the valley portion of the profile is placed under tension and the crown under compression. In the case of multispan sheeting the condition reverses over the intermediate purlin. With wind suction loading, these deflection and stress patterns are reversed.

With profiled metal sheets there is little risk of failure in tension, with a rather greater risk of failure where compressive forces are concentrated This fact is sometimes accounted for in design by the introduction of additional mini-stiffener ribs, roll-formed into the profile.

Apart from the rigidity obtained from the profile, the strength of the metal is marginally increased by the cold working of the metal during the roll-forming process.

With fibre-cement profiled sheeting the cement matrix is weak in tension and strong in compression. Tension load strength therefore depends on the tensile strength and integrity of the fibres used to reinforce the cement matrix.

1.3.1.1 When a suitable thermal insulant is introduced by foaming between a profiled metal weather sheet and a metal lining under controlled conditions, the resulting composite sheet or panel will have a far greater resistance to deflection than the sum of the strengths of the individual parts.

If however, due to abnormal deflection produced by excessive loads or differential thermal expansion, or for any other reason, the core material or its bond with the cover sheets should fail, the strength of such a unit would revert to that of the separate components.

1.3.1.2 In semi-composite construction where the individual elements are brought together on site without bonding, the performance of the outer sheeting and of the lining panel must be considered separately.

1.3.1.3 To simplify the process of selecting profiled metal roofing/cladding, it is usual to refer to manufacturer's load/span tables which give for each profile and metal thickness:

- the maximum recommended span (for single and often for multiple span applications);
- the relevant load and the resulting deflection.

It is the duty of the specifier to determine the appropriate parameters of load/deflection. Unless or until agreement is reached on a uniform approach to the presentation of these data, it is necessary to exercise caution in interpretation as follows.

- Check precisely what is included in the load figures.
- On what basis is the deflection assessed? This is given as a fraction of the span and may vary from a high limit $L/90$ (1/90th of span), to a low limit of say $L/200$ (1/200th of span).
- Is any provision made for wind load and if so, on what basis?
- Note all loads for which an allowance should be made over and above those included in the tables (specific concentrated loads, drifting snow loads, etc.) and refer to the manufacturer for guidance if necessary.

Load span tables for fibre-cement sheeting are usually limited to quoting a maximum span for so called 'Normal Exposure' conditions. The fibre-cement manufacturer should be consulted to establish what design wind load can be accepted for any particular building layout and location.

No simple selection procedure can be comprehensive and thought should be given to such features as parapets, open canopies, abnormal sheet overhang at eaves and areas designated as subject to 'local' pressure coefficients in the wind loading codes BS 6399: Part 2 and CP3: Chapter V: Part 2, particularly when locating rooflights and other integrated lightweight auxiliary items, e.g. soaker sheets, etc.

At the time of writing, both BS 6399 and the older CP3 wind loading codes are available, and they can give significantly different design values for some types of building especially on local zones. Designers for any site should check if the Local Building Control Authority has decided to use BS 6399 or if CP3 is still to be applied.

1.3.2 Openings in roofing/vertical cladding

When an opening, which may result in weakening of the sheet, is to be formed through profiled roofing and cladding, due consideration should be given to the addition of trimmers between the appropriate purlins or rails (and any necessary cross-trimming) to ensure adequate support for the sheet.

1.3.3 Purlin/side rail restraint

The primary function of profiled sheet metal roofing and cladding is to provide a weather resisting envelope. Nevertheless, when profiled metal sheeting is secured to the supporting members using rigid fixings it will automatically impart a measure of restraint. This must not be construed as relieving the designer of the framework or the contractor undertaking its erection of responsibility for the provision of suitable anti-sag bars or diagonal bracing.

Primary fixings in certain proprietary secret-fix or standing seam sheeting systems are not secured by through-fixed fasteners but may be free to slide in the longitudinal direction of the sheeting. Such systems should not be relied upon to provide lateral support to purlins or spacer members.

Likewise fibre-cement sheeting will stiffen a structure when correctly fixed, however, any such stiffening effect should not be included in the design calculations.

Purlin and rail sizes should be determined by the structural designer, by reference to the profile manufacturer's tables, to satisfy the loads discussed in section 1.2 in relation to spacing and span sheet configurations.

The performance of fasteners in purlins and rails is dependent upon profile, thickness and yield strength. Satisfactory performance can be achieved using 1.2 mm thickness steel with a minimum yield strength of 350 N/mm² or 1.5 mm thickness with a minimum yield strength of 280 N/mm². For zed spacers the minimum thickness should be 1.5 mm to provide satisfactory stability and strength. For proprietary systems, designers should consult the manufacturer's recommendations. As the thickness or yield strength is reduced the fastener pull-out and stripping torque reduces making it essential that screw guns are set with correct depth locators or torque controls.

Designers should always obtain appropriate data for the fasteners. Any purlin, rail or spacer to which roof sheeting and cladding is to be fixed should provide a bearing surface as required by the material manufacturer's recommendations, particularly at end laps.

It is the responsibility of the structural frame contractor to ensure that purlins and rails are aligned true and parallel, that they are correctly connected to the main structure and that all necessary anti-sag rods and bracing are fitted. All structural frames and their members should comply with the relevant Code of Practice and be erected within the specified dimensional tolerances.

1.3.4 Stressed skin performance

When profiled metal sheeting/cladding is by design, to be incorporated into the building as a structural component, with the intention of resisting certain of the loads to which the building will be exposed, the technique is usually termed 'stressed skin' design. Probably the most common application of stressed skin design is in resisting the effects of wind load, by providing lateral stability in the plane of the cladding.

Whether the framework is to be considered as 'flexible' or as providing partial restraint acting in conjunction with the cladding, the division of responsibility for design and performance, between designer of the structural frame and supplier of the cladding material, needs to be clearly defined prior to contract.

Apart from establishing the suitability of the profiled metal sheeting in terms of mechanical performance, careful consideration must be given to the loads to be transmitted by the fixings to the supporting frame and the reactions between fasteners and sheeting which, by creating elongated fixing holes, may prejudice the weather resistance of the envelope.

The effects of localized weakening of the cladding system by the introduction of rooflights, ventilators, etc. must be evaluated and attention paid at all stages to maintaining stability during construction (see 1.2.4 Wind Load).

This form of construction would warrant professional advice if, at any time after completion, it should become necessary to remove sheeting, which may put at risk the stability of the structure.

1.4 THERMAL MOVEMENT

1.4.1 Profiled sheeting

The commonly used materials of construction expand with a rise in temperature and contract with temperature reduction and although the dimensional changes occur in all directions, it is the effect on length which is of primary concern to users of profiled sheets for roofing and cladding.

The change in length is proportional to the original length and also to temperature change and is calculated using the 'coefficient of linear expansion per °C' for the material under consideration.

1.4.2 Temperature range

The increasing use of thermal insulation beneath profiled sheeting has considerably reduced the influence of internal building temperature on the temperature of the cladding. As a result, the sheeting undergoes greater changes in temperature under the combined influence of outdoor air temperature and solar radiation than would otherwise occur on poorly insulated buildings.

The effect of solar radiation is modified by surface characteristics to the extent that a dark (brown or black) roof finish may attain a temperature 30°C above that of a light coloured surface, when both are backed by insulation and subject to the same exposure.

Table 1.1 Typical thermal movement values for materials relevant to this Guide

Material and coating colour (if any)	Coefficient of expansion $\times 10^{-6}$	Typical temperature range (°C)		Overall movement (mm/m)	Movement about ambient (mm/m) (ambient = 5°C)	
Aluminium mill or light	23–24	−10	+50	1.38	−0.35	+1.04
Aluminium dark	23	−10	+70	1.84	−0.35	+1.50
Steel light	12	−10	+45	0.66	−0.18	+0.48
Steel dark	12	−10	+70	0.96	−0.18	+0.78
Stainless steel	14	−10	+70	1.0	−0.21	+0.91
Rigid PVC (clear)	68	−10	+30	2.7	−1.02	+4.42
GRP	25	−10	+40	1.3	−0.38	+1.63
Polycarbonate	67	−10	+30	2.7	−1.01	+4.36
Fibre-cement	8	−10	+40	0.40	−0.12	+0.52

The ambient temperature during sheet installation has been assumed to be 5°C.
If the sheeting is installed during very cold weather the temperature range base should be decreased to −20°C.

In the United Kingdom, maximum temperatures will occur for periods of several hours at a time on roofs of a dark colour, typically in late June or early July. For vertical cladding the corresponding temperatures will be lower, peaking on elevations from east through south to west during the day.

The greatest 24 hourly temperature change occurs in summer and is generally regarded as the condition to be considered in relation to thermal movement in cladding.

1.4.3 Accommodating thermal movement

1.4.3.1   The ability of profiled metal sheet to accommodate lateral movement by its concertina-like flexibility, means that no special provision is needed for thermal expansion in the direction of sheet width, but a conflict of interests arises where longitudinal thermal movement is concerned.

A desire to limit the number of end laps and therefore the number of potential leaks, favours the use of sheeting in the greatest available lengths but from Table 1.1 it is seen that sheet length directly affects the amount of movement to be accommodated. For example, the notional linear expansion of an aluminium sheet 10 m in length and finished on the exposed face in a dark colour would be 18.5 mm.

It should equally be understood that if two or more sheets of similar material are lapped and fastened together preventing relative movement making up the same aggregate length of 10 m, the same overall expansion of 18.5 mm will result.

If the sheet is fully restrained by fixings at one end into say, an inflexible ridge purlin, the total movement should be allowed to take place at the other end and proportionally at intermediate purlins. Alternatively, if

it were possible for the sheet to be restrained by fixings into an inflexible purlin central in the sheet length, then half the total movement would need to be accommodated at each end and proportionally less at each intervening purlin.

1.4.3.2   In practice thermal movement will normally be accommodated by deflection of the sheet between purlins, flexure of metal spacers, and by the flexing of purlins.

Field experience indicates that no special provision need be made for thermal expansion in profiled steel sheet of total length less than 25 m, whether finished in light or dark colours. Secret-fix sheeting can be used in single lengths exceeding 25 m, and will require adequate provision for thermal expansion.

For dark coloured aluminium sheets, a maximum sheet length of 9 m is recommended and for a light coloured finish, 11 m. Provision should be made for movement at end laps in accordance with the manufacturer's recommendations.

If the overlap at an expansion joint needs to be sealed, the sealant must be capable of allowing the relative movement of the sheets to take place without detriment to the seal. The use of high temperature grease (lithium based) is recommended by some aluminium profile sheet manufacturers rather than a sealant with high adhesion.

The use of curved sheets at eaves and apex requires careful consideration because of thermal movement if the problems of 'pull-out', 'pull-over' and 'washer inversion' at fixings or excessive over-turning moments at purlins are to be avoided. An eaves design incorporating a concealed gutter will at least give an opportunity to accommodate some movement of the sheeting, but the potential for thermal expansion in the total length from gutter to gutter will inevitably be too great to be ignored.

1.4.3.3 Where composite panel systems are employed thermal movement must be considered in relation to the weathersheet, foamed insulating core and lining sheet.

Changes in temperature differential from one surface to the other produce differential expansion and a tendency to bowing which, in summer, will increase the tensile and shear forces on fasteners. Fasteners for composite panels are therefore purpose designed and have to comply with industry tests. Only fasteners which have the full approval of the composite panel manufacturer should be selected.

Provision for thermal movement must retain continuity of insulating effect and integrity of vapour control at the lining, if interstitial condensation is to be avoided.

Site assembled semi-composite panel systems do not pose the same problem as each element can be dealt with independently and no shear forces are transmitted to the insulant.

The comments relating to roof construction and thermal movement apply also to wall cladding, particularly where profiled metal sheets are fixed horizontally in long lengths and/or to curved corner units.

1.4.3.4 Forces produced by sheeting restraint

The amount of linear expansion, especially in a relatively short sheet, may seem insignificant but if the sheet is fully restrained at both ends and without deflection, considerable forces are developed.

Whilst the force developed in the direction of the length of a fully restrained sheet is easily calculated as the product of the transverse cross-sectional area of the sheet, the modulus of elasticity for the material, its coefficient of linear expansion and the change in temperature causing the expansion, interpretation of the result is far from straightforward.

It is for this reason that the freedom given by the features mentioned in 1.5.3.2 is of importance and must be given due consideration by the designer of the structural frame.

1.4.3.5 Thermal movement in rooflights

From the data set out in Table 1.1, it will be noted that polycarbonate (PC) and rigid PVC expand to a far greater extent per °C of temperature change, than any of the metals or GRP.

Thus it is sound practice, for rooflights constructed from polycarbonate and PVC, to drill fixing holes 4 mm larger in diameter than the shank of the primary fastener (see Rooflights – section 6.3.3). The fasteners should be centred in the fixing holes to allow freedom for expansion or contraction of the rooflights. Manufacturer's recommendations should be sought on choice of fastener, fixing procedure and the provision for thermal expansion.

The normal deflection of double skin rooflights fabricated from polycarbonate and PVC is likely to be exceeded, due to the differential expansion of the two skins, when the outdoor temperature is significantly higher than the indoor condition. This may result in ponding on rooflights installed in low pitched roofs.

1.4.3.6 Movement joints in fibre-cement cladding

Fibre-cement sheeting requires movement joints for reasons which are explained in Chapter 5.

1.5 DETERIORATION – MATERIAL INCOMPATIBILITY

1.5.1 Material incompatibility

Materials comprising the building envelope, although required to meet particular design criteria, must not be considered in isolation. A component, which in functional terms seems ideal, may prove unsatisfactory in service because of an unforseen incompatibility with other material or substance in contact with it.

1.5.2 Bimetallic corrosion

Bimetallic corrosion occurs when dissimilar metals are in direct contact in the presence of an electrolyte. One of the metals will form the anode the other the cathode, whilst moisture affected by atmospheric pollution acts as an electrolyte. A galvanic cell is thus produced and a minute electric current causes corrosion of the anodic metal whilst the cathode is protected.

Base metals are anodic, noble metals cathodic and although the sequence may be modified depending on the electrolyte, a guide as to the risk of corrosion is given by the relative positions of the metals in the 'electromotive series'.

1.5.2.1 Example

At temperatures below about 65/70°C zinc is anodic to steel, the zinc coating (as in galvanizing) is therefore slowly sacrificed to give protection to the steel for a period. (At temperatures above 65/70°C an unusual inversion takes place and the zinc is protected at the expense of the steel.)

1.5.2.2 The following reactions are liable to occur and the pairings should be avoided in the presence of moisture:

- copper and copper alloys corrode cast iron, mild steel, cadmium plated and galvanized steel, zinc and aluminium;

- mild steel corrodes galvanized steel and aluminium (in industrial and marine environments);
- lead corrodes aluminium (especially in industrial and marine environments) and cast iron;
- aluminium is liable to corrode galvanizing, but when it is left in contact with the steel substrate the aluminium will oxidize;
- aluminium alloys with appreciable copper content corrode aluminium alloys containing zinc, especially in marine and industrial environments.

1.5.2.3 The behaviour of aluminium and stainless steel in contact depends largely upon the passivity or oxide-film forming ability of the stainless alloy. In this respect the austenitic grades containing generally 17–26% chromium and 8–22% nickel are unlikely to cause problems with aluminium except in the most unfavourable environmental conditions. It is the aluminium sheet which is likely to suffer rather than the stainless steel fastener.

Cadmium plating, albeit being phased out, is used to provide protection to small steel components but being anodic, acts sacrificially. Nickel is by comparison cathodic and when deposited on steel relies on soundness of coating to give protection. Corrosion will be promoted at any discontinuity in the coating.

1.5.3 Avoiding bimetallic contact

If the use of incompatible materials cannot be avoided, bimetallic contact may be reduced by introducing a suitable separating material possessing the characteristics of an electrical insulator.

1.5.3.1 Separation of aluminium cladding from steel purlins and rails or galvanized spacers, can be achieved by the use of an inert electrical insulating material such as PVC tape, or suitable paint, e.g. zinc chromate, iron oxide or bitumastic. Lead-base paints and materials containing free chlorine should be avoided.

1.5.3.2 The question of material incompatibility does not arise when steel cladding is used in conjunction with steel framing.

The use of zinc as a sacrificial coating to give protection to a steel substrate is referred to elsewhere, but where such material used as a base is then given a durable, moisture impermeable, decorative finish, the action of the anodic zinc only comes into play if the surface coating is penetrated, e.g. at cut edges.

The risk of inverse bimetallic corrosion should be avoided by refraining from the use of dark coloured surface finishes. The available range of lighter shades makes it possible to limit the surface temperature of cladding to a level (under summer conditions in the UK and with insulation beneath) below that at which the phenomenon is known to occur.

The alternative of painting the exposed edge of each overlapping top sheet would add appreciably to fixing costs and if required must be clearly specified.

1.5.3.3 Where fasteners are concerned a degree of isolation is given by the resilient, non-metallic sealing washer and separation of screw and sheeting may be provided by electroplating or an epoxide or polymer coating applied to the fastener.

Alternatively, the use of stainless steel fasteners with suitable sealing washers is now a widely accepted solution, despite the cost implication.

1.5.4 Surface protection

1.5.4.1 Steel drilling swarf will promote rust staining and swarf is liable to damage surface finishes, so it should be removed with care as work proceeds, to avoid abrasion and soiling.

Cement and plaster should be removed whilst wet if scratching and discoloration are to be prevented.

Rain water run-off from roofs affected by moss or algae will be acidic and capable of attacking unprotected metal.

Similar effects will be produced by water from surfaces of copper or lead. The problem in either case may be tackled at source by removing vegetation or painting with bitumastic paint respectively, or the recipient metal surfaces may be similarly painted.

Deposits of industrial grime can soon disfigure the surface of cladding. This staining will show with greater effect on textured finished or on the natural oxidized surface of aluminium.

1.5.4.2 Any metals in contact with fibre-cement or rain water run-off may be severely attacked by the alkali dissolved out of the cement. Fasteners should therefore have alkali resistant coatings.

Fibre-cement is attacked by some types of chemicals found in a polluted industrial environment and the manufacturer's advice should be obtained before ordering. Guidance on durability and compatibility is given in Chapter 5.

1.5.4.3 Changes in appearance will also take place in time under such influences as surface temperature, UV light, thermal movement, local environmental contamination, etc., the effect depending upon the nature of the surface finish.

For this reason, manufacturers should be consulted on the durability of available finishes and the anticipated 'period to first (major) maintenance'. Alternatively, guidance is provided on the wide range of materials in generic terms, in BS 5427.

It is recommended that designers and specifiers inform their clients of all suppliers' recommendations on the regular programme of maintenance required for all forms of cladding. This is so that the recommendations are part of an adequate preventative maintenance programme.

1.5.5 Latent deterioration

1.5.5.1   Serious thought should be given at the design stage to the risk of deterioration of internal surfaces, including those that are concealed by linings and insulation.

Section 1.9, examines the topic of condensation, the insidious nature of which should not be underestimated. The risks should be reviewed in the light of material or processes to be employed on the premises, which might promote corrosion.

1.5.5.2   The use of timber spacers and purlins is now largely restricted to the agricultural market. Specifiers should be aware that some timber preservatives can cause rapid corrosion of metal fasteners and metal sheeting. The type of preservative used should therefore be checked for suitability in advance with the profile and fastener manufacturers.

1.6 BUILDING REGULATIONS – FIRE

1.6.1 General

Part B of Schedule 1 to the Building Regulations 1985 seeks to ensure that buildings are so constructed that in case of fire:

(a) the occupants are given a safe means of escape;
(b) the spread of fire within and between buildings is minimized;
(c) collapse will be delayed to permit evacuation and avoid contributing to rapid extension of the fire.

1.6.1.2   The requirements for limiting the spread of fire within and between buildings are met by: compartmentation; standards of non-combustibility and fire resistance for external walls; imposing limits on surface materials of walls and ceilings; standards for flame spread and resistance to penetration by fire of roofs; and the use of fire stops and cavity barriers.

1.6.1.3   The requirement for stability in fire is met by setting standards of fire resistance for the structural elements (floors, frames and load bearing walls and roofs).

1.6.2 Profiled sheeting and cladding

Three types of fire performance are currently mentioned in the Building Regulations which relate to sheeting and cladding, these may be summarized as follows:

1. Roof coverings – resistance to external fire being spread from adjoining premises [designated by two letters, e.g. AA, AB or CA, etc., indicating the time to penetration and the distance spread of flame along their external surface. Test method BS 476: Part 3: 1958].
2. Walls and other surfaces – combustibility, class of surface spread of flame, fire propagation performance, ignitability, etc. [designations might include non-combustible, Class 1 Surface Spread of Flame or Class O in accordance with Building Regulations; test methods include BS 476: Parts 4, 6 and 7].
3. Walls may need a period of fire resistance, with stability, integrity and insulation attributes [designations may be given as ½ hour, 1 hour, up to 6 hours; the period is determined by test in accordance with BS 476: Parts 20, 21, 22 or 23].

1.6.2.1   Under the Building Regulations, roofs are not required to provide fire resistance but should resist penetration by fire from the outside and limit the spread of flame over their surfaces.

BS 476: Part 3: 1958, External fire exposure roof test, specifies tests on a specimen of 'roof construction' to include all roof covering constituents, any membrane or insulation fixed over-purlin, decking and supports but excluding any under-purlin lining or ceiling.

Note: This standard has been withdrawn by the British Standards Institution but continues to be cited in the Building Regulations. The relevant test method has therefore been included as an annex in BS 5427: 1996.

In Approved Document B of the Building Regulations (AD B), Appendix A, Table A5 supplies notional designations of roof coverings. Pitched roofs covered with self-supporting profiled sheet of galvanized steel, aluminium, fibre reinforced cement, or pre-painted (coil coated) steel or aluminium with a plastisol or PVF_2 coating as single skin without underlay or with underlay or plasterboard, fibre insulating board, or woodwool slab on a supporting structure of steel, concrete or timber are given the notional designation AA. This is the highest designation.

Similarly pitched roofs covered with self supporting sheet of galvanized steel, aluminium, fibre-cement or pre-painted (coil coated) steel or aluminium with a plastisol or PVF_2 coating as a double skin without interlayer, or with an interlayer of resin bonded glassfibre, mineral wool slab, polystyrene, or polyurethane on a structure of steel, concrete or timber are given the notional designation AA. In terms of separation distances to the relevant boundary wall, which may be a notional boundary, there are no restrictions on the use of roof coverings designated AA, AB or AC.

1.6.2.2   In the past there was a clear distinction between materials classified non-combustible when tested to BS 476: Part 4 and all others. A new term 'materials of limited combustibility' has been introduced into building control legislation and relates to materials assessed by reference to BS 476: Part 11: 1982 (1988), Method of assessing heat emission from building materials.

AD B, Appendix A, Table A6 describes the use of and definitions of non-combustible materials.

AD B, Appendix A, Table A7 contains definitions of, and details where, materials of limited combustibility may be used, including reference to certain insulation linings, roof coverings and cavity barriers.

AD B, Appendix A, Table A8 provides information on the performance rating of some generic materials which rate Class O or Class 3. Products listed under Class O also meet Class 1 Surface Spread of Flame.

The Class O designation and fire propagation index are used to impose limitations on external cladding according to the height of the building and its distance from the relevant boundary (Tables 1.2 to 6.2 of Approved Document B2/3/4). The fire propagation index is determined by BS 476: Part 6: 1989.

1.6.2.3   BS 476: Parts 20–23 deal with test methods for the fire resistance of elements of a construction and the contribution of components. These Parts of BS 476 are concerned with the fire resistance of complete elements of construction and provide for tests on walls and partitions (load bearing or non-load bearing), flat roofs, suspended ceilings protecting steel beams and floors and ceiling membranes which provide a separating function and in Part 23, the contribution of components to the fire resistance of a structure.

During tests, observations are made to determine the 'stability', 'integrity' and 'insulation effect' of the element with one face exposed to furnace temperature under controlled conditions.

Failure is generally deemed to have occurred in respect of:

- stability – when collapse takes place (even of non-load bearing elements);

- integrity – when cracks or other openings occur through which flame or hot gases pass, igniting a sample pad at a prescribed distance from the unheated face;
- insulation effect – when the unheated surface increases in temperature by a specified extent above the initial temperature.

In general, the fire resistance of an element is the time in minutes from commencement of the test until failure occurs under any one of the three criteria or, if no failure occurs, until the test is terminated.

1.6.2.4   In terms of internal fire spread where cavity barriers are required, the provisions of maximum dimensions of cavities in non-domestic buildings do not apply to the cavity between double skinned or profiled insulated roof sheeting, if the sheeting is a material of limited combustibility and both surfaces of the insulation layer have a surface spread of flame of at least Class 1 or O and make contact with the inner and outer skins of the cladding.

1.6.3   CEN fire regulations

1.6.3.1   The European Committee for Standardisation (CEN) works under the direction of the European Commission (EC) to produce under the Construction Products Directive (CPD, 89/106/EC) harmonized standards for construction materials which will remove barriers to trade in the politically defined Single Market by describing the fitness criteria for materials which will permit products to carry the CE mark which signifies 'proof of fitness to be on the market of every member state'. CEN is developing Product Standards and Methods of Test which will be published and applied throughout the Member States (MS) and will replace British Standards in UK. BS ENs may be described as performance standards, whereas the British Standards were more prescriptive.

1.6.3.2   New methods of test for reaction to fire and external fire performance of roofs, together with measures to harmonize furnace test conditions in MS for fire resistance tests, are being developed in CEN. It seems probable that materials subject to a Reaction to Fire test will be classified from A1, A2 indicating non-combustibility, through B to F for unclassifiable.

For external fire exposure of roofs a Method of Test has been drafted and may be imposed which contains three different test methods. It is probable that a list of roofing products will be authorized by the EC which will be exempt from testing but only when used in the specified typical end use applications. This list may be similar in form to the list of products to be considered as Reaction to Fire Class A established by the Decision 94/611/EC without the need for testing.

Slates and tiles, profiled metal, fibre cement and translucent sheets in GRP, PC and PVC, fully supported metal and all other forms of roof covering will all come under the external fire exposure performance requirements. The parameters described for the EC by the Fire Regulators Group for roofs exposed to external fire are:

- fire penetration;
- fire spread across the external surface of the roof;
- fire spread within the roof;
- production of flaming droplets or debris from the exposed surface.

It is clear that the list of 'deemed to satisfy' products which do not need to be tested will only include products (in specified end use applications) which can be shown to have adequate resistance to these parameters to the satisfaction of the European Commission's Regulators.

No direct correspondence between BS 476 Surface Spread of Flame classes and the new Reaction to Fire classes can be expected. Profiled metal pre-painted (coil coated) with a PVC or PVF_2 coating, composite/ sandwich panels with foam insulation, GRP, PC and PVC rooflight materials may be graded in classes yet to be identified to meet UK regulations.

1.6.3.3 Fire resistance of walls

It is anticipated that the changes being made to harmonize furnace test conditions may alter the thermal exposure in some furnaces and for some specimens. It remains to be seen how this may alter the fire resistance times required to meet national regulations.

1.6.3.4 Transitional arrangements

It is anticipated that after a transitional period, which may last up to ten years, all materials will need to be tested under the appropriate CEN test standard for Fire Resistance or Reaction to Fire. From the introduction of CEN test standards, i.e. BS ENs at their implementation date, all new products to the market must be tested to these BS EN standards.

Note: The BS 476 test methods will only be acceptable within the transition period for products tested before the implementation date which may be 2000.

1.7 FIRE PERFORMANCE

1.7.1 Fire requirements for walls

The Building Regulations AD B specifies whether an element of structure is to be tested by exposure of one side or both sides separately, giving the period of fire resistance demanded under each of the three headings. It also lays down the minimum periods of fire resistance for buildings according to purpose group and size.

Metal cladding alone would fail to meet the criteria for fire resistance and under the terminology of AD B, appendix L would be an 'unprotected area'. Used as part of an element of structure with suitable lining materials it might achieve a fire resistance rating.

Depending upon purpose group and distance from a relevant boundary AD B4, diagram 44, allows for a certain extent of external wall to be unprotected, allowing the use of metal cladding without specific fire resistance.

1.7.2 Updating fire requirements

Owing to the ongoing programme for the revision of British Standards it is inevitable that some will be withdrawn, although they may remain cited in the Building Regulations Approved Documents. New Standards will be produced and may not appear in the Approved Documents. It is recommended that designers and users periodically review Standards and Approved Documents for current status.

1.7.3 Rooflights

1.7.3.1 This subject is described in detail in Chapter 6, paragraphs 6.5 *et seq.*, as at October 1998.

The use of rooflights of plastic material with a lower surface of not less than Class 3 surface spread of flame (to BS 476: Part 7: 1987) are permitted, but made subject to limitations of area per rooflight, area as a percentage of floor area and distance of separation by material of limited combustibility, relative to purpose group.

The rules are summarized in AD B section 14, paragraphs 14.4 *et seq.*, diagram 52 and Tables 16 and 17.

1.7.4 Loss Prevention Council

The Loss Prevention Council (LPC) is funded by the insurance companies to provide advice on fire matters and independent testing of manufacturers' products. LPC have established a number of tests, reference LPS1181, which measure the fire performance of a sample roof and wall assembly. This type of test procedure is not included in corresponding British or European standards at this time.

1.7.5 FM approval

Factory Mutual is a large American organization with world wide interests in reducing loss from building failures. Although similar to LPC, FM approval of a product will include limitations on how it is installed with other products. These limitations may include the make or manufacturer of the fasteners which have to be used.

1.8 BUILDING REGULATIONS – VENTILATION

1.8.1 Approved Document F

The title of Approved Document F to the Building Regulations is *Ventilation 1995 edition.* It is anticipated that the Building Standards (Scotland) Regulations and Building Regulations (Northern Ireland) will also be revised to include similar requirements as these Approved Documents.

The revised documents are intended to reduce the overall heat loss through the building envelope and reduce the risk of condensation within the structure. Reference is made within the Approved Documents to the revised Building Research Establishment report *Thermal insulation: Avoiding risks – Report No. BR 262.*

The requirements for ventilation given in Approved Document F2 state that:

'the requirement will be met if condensation in a roof and in the spaces above insulated ceilings is limited so that under normal circumstances

a. the thermal performance of the insulating materials and

b. the structural performance of the roof construction will not be substantially and permanently reduced.'

Methods to minimize the risk of condensation are included in:

● Chapter 3 – Built-up Metal Roofing Systems;
● Chapter 4 – Composite Panels;
● Chapter 5 – Fibre-cement Sheeting.

1.8.2 Internal environment

The risks of condensation building up sufficiently to cause long term damage will vary with the conditions inside the building. These conditions are dependant on the number of people present, manufacturing processes and heating conditions inside the building. It is important that the designer of a building assesses the risk at an early stage and establishes the humidity level so that appropriate systems and materials are used. It is essential that the roofing contractor is provided with this information.

1.9 REDUCING THE RISK OF CONDENSATION

1.9.1 The role of the designer and contractor

1.9.1.1 Condensation is not a new phenomenon, but it has become more important during recent years. This is a direct result of energy conservation measures such as increasing the amount of thermal insulation, and of the demand for warmer working conditions in factories and offices. It is also a consequence of improved construction standards, because better fitting flashings, doors and windows, reduce the amount of natural ventilation within buildings. The external metal skin of a roof or wall can become very cold in winter; this may result in condensation forming on the inner surface of the metal, which can be harmful unless the correct precautions are taken in the design and construction.

1.9.1.2 The designer should be aware of the intended use of the building, and should take account of this in arriving at a specification. Where the use of the building may change during its lifetime, the designer should ensure that the owner is aware of the possible risks associated with a change of use.

1.9.1.3 Sensible planning to eliminate or greatly reduce risks includes design of appropriate details, and the selection of suitable materials of good quality. The suitability of the materials should be judged in terms of durability as well as performance. *The condensation protection must last at least for the anticipated life of the roof,* it should not be jeopardized by corrosion, decay or disintegration.

1.9.1.4 It is important that technical representatives and salesmen, surveyors, estimators and contract managers, should be completely familiar with the causes of condensation, and how it can be reduced or prevented. This essential knowledge enables them to advise customers of potential problems before they arise, and before contracts are signed. It is equally important that operatives are properly trained, in the purpose, function, and in the correct installation methods for vapour control features.

1.9.1.5 Uninsulated single skin metal and plastic rooflights are prone to severe condensation in adverse weather conditions.

1.9.2 Air, moisture and condensation

1.9.2.1   All air holds water in the form of vapour. The amount which is held is influenced by the local environment and processes; however, the maximum amount is governed by the temperature. If air is heated, it can hold more water vapour (this is why most drying equipment uses heat or a flow of air – or both). For measurement, calculation and control purposes, the moisture content of air is expressed as kilogrammes of water per kilogramme of dry air. When air contains the maximum amount of water possible, at that temperature, it is said to be *saturated.*

1.9.2.2   Most air is neither dry nor saturated, and it is necessary to have some means of comparison. The simplest comparison is the percentage saturation; this is the number of kilogrammes of water per kilogramme of dry air, expressed as a percentage of the maximum possible at that temperature. This is not quite the same as the relative humidity, although the numerical values of the two functions are virtually identical. When water is added to air as a vapour, it creates a vapour pressure; more water makes more pressure. Relative humidity is the ratio of vapour pressure in air, expressed as a percentage of the vapour pressure in saturated air at the same temperature.

1.9.2.3   If a sample of non-saturated air is cooled progressively, its percentage saturation (and relative humidity) increases until the temperature is reached at which the air is saturated. This temperature is called the dew point. If a sample of air is cooled below its dew point, it must release some water vapour as dew or condensation, because air cannot be more than 100% saturated.

1.9.2.4   It will be appreciated that the dew point is a function of temperature and percentage saturation (or relative humidity). For example, an office may operate at a temperature of 20°C and a relative humidity of 50%; the dew point would be at approximately 9°C. Some human activities, such as washing, bathing and cooking, put more vapour into the air and increase the relative humidity. In the above example, an increase to 70% relative humidity would raise the dew point to approximately 14°C. If the windows were single glazed, and the outdoor temperature between 9°C and 14°C, the latter case would result in condensation on the windows, but this would not have occurred in the former case.

1.9.2.5   The temperature of outer sheets can fall a few degrees below outside air temperature on nights without cloud cover because of clear night sky radiation. This lower temperature increases the possibility of condensation occurring and the condensation could freeze if outside conditions were severe.

1.9.2.6   So, it is possible to reduce condensation in two ways. First, ensure that surfaces in contact with humid air are kept warm, i.e. above the dew point of the air. Second, keep the relative humidity as low as possible so that the dew point is also low.

1.9.2.7   The use of thermal insulation is one way in which internal surfaces can be kept warm. An obvious example is the use of double glazing; this is affected by condensation far less frequently than single glazing, because the air space keeps the inner surface warmer. *However, thermal insulation alone is not the complete answer to minimizing condensation;* it may keep the inner surfaces warm, but temperatures will be lower at various positions within the construction. Again double glazing provides a useful example; poorly sealed systems can suffer condensation within the cavity, because humid air is able to by-pass the warm inner pane and access the cold inner surface of the external pane. Condensation is not so immediately visible in insulated roof constructions, but can occur just as easily if humid air can by-pass the warm lining and approach the cold external roof sheets.

1.9.2.8   The most effective way of reducing excessive humidity, is by the use of ventilation, i.e. by moving the moist air outside the building and replacing with drier air. This can be natural fixed ventilation, as with ventilating ridges over continuously humid activities, or sophisticated electronically controlled fans and heaters, such as are used in modern swimming halls.

1.9.3 Movement of moisture through the construction

1.9.3.1   Humid air, or vapour, can pass through construction materials as a result of simple air movements, or through diffusion under vapour pressure.

1.9.3.2   Air within buildings is constantly moving due to convection, draughts, movements of people or equipment, etc. The air will move into the construction at unsealed joints in the lining, unsealed openings around pipes or service ducts, badly fitting frames at doors and windows, etc. As vapour can pass through such openings in great volumes, it is important that they are closed and sealed as effectively as possible.

1.9.3.3   Diffusion takes place without the need for air movement; it is the result of differential vapour pressure. When water enters air, as a vapour, the pressure of the air increases; if the amount of vapour increases, the air pressure also increases. Usually, the air within a building contains more vapour than the outside air, so its pressure is greater. This means that there is usually a positive pressure differential, trying to push moisture vapour out of the building. No materials are totally impervious to moisture vapour (although some are very nearly so), so the pressure differential pushes humid air from the inside of the building into the construction. In winter conditions, the temperature falls as the vapour moves closer to the outside sheets; if the temperature falls to the dew point of vapour, condensation will start to form – this is known as interstitial condensation. It is not possible totally to prevent the diffusion of vapour, but its effect can be minimized by the correct choice of materials, good detail and design, and high quality workmanship.

1.10   VAPOUR CONTROL LAYERS

1.10.1   The use of a VCL

1.10.1.1   These used to be called vapour barriers, but it is now acknowledged that a complete barrier to vapour is an impossibility. A vapour control layer (VCL) is intended to be a sufficient barrier against the diffusion of vapour, taking account of the use of the building. The VCL over a swimming hall, textile mill or food processing area, will need to offer greater resistance than that over a dry goods warehouse, but it is important to note that BRE report 262 states that built-up insulated roofing should only be used where it is possible to fit a vapour control layer on the warm side of the insulation. Building Regulation – Approved Document F2 specifically mentions BRE report 262 as a source of detailed guidance.

1.10.1.2   The purpose of the vapour control layer is to prevent vapour from reaching the cooler parts of the construction. It is therefore vital to its function that it should be fitted at, or close to, the warm side of the insulation – that means close to the lining of heated buildings (in other parts of the world, different conditions may prevail).

1.10.1.3   It is sometimes argued that vapour pressures are very small, and therefore easy to resist. However, it must be appreciated that although the pressures are small, they can persist continuously over very long periods. The BS 5250 recommendations for the basis of assessing condensation risk, are internal conditions of 15°C and 65% relative humidity and average external winter conditions of 5°C and 95% RH for a continuous 60 day period. These are helpful for

'average' conditions, but are far less severe than the conditions which would apply in swimming halls or similar high humidity environments.

1.10.2   Designers' responsibilities

1.10.2.1   Designers should attempt to produce details which are capable of being sealed adequately. Both designers and contractors should appreciate that some vapour control materials are more efficient than others. Quality and performance, not price, must be the overriding consideration. Site operatives should be trained in the use of the specified product, to ensure that it is used correctly.

1.10.2.2   With typical built-up metal cladding, i.e. 'hybrid construction' (as defined in BS 5427) the liner panels with all side and end laps sealed and washered fasteners used to fix the liner panel may be classed as the VCL except for high internal humidity conditions, see below. The seals may be mastic sealants within the laps, or adhesive tapes fixed over the laps from above. Both methods are dependant on the standard of workmanship and the strength and condition of the liner edge. Seals which are omitted, or fail to adhere, can negate the whole vapour control method. Tapes offer better prospects for inspection during construction, but may be more vulnerable later as the building experiences movements due to wind, snow and changes of temperature. Any fasteners which penetrate the liner should have compressible washers to maintain the vapour seal.

1.10.2.3   In the past, ordinary 500 gauge polyethylene sheeting has sometimes been used as a vapour control layer. The tear and vapour resistance of this sheeting is so low that it really should not be considered as a viable option when designing or choosing vapour control systems.

1.10.2.4   The use of thicker gauges (e.g. reinforced 1000 gauge or more) can be a help, but the material must be 'virgin' because recycled products can have minute pores or pinholes which drastically reduce their vapour resistance. Prospective users should be aware that ordinary polyethylene may not be an appropriate material for use as a vapour control layer in some constructions, as it can become thermally unstable at temperatures as low as 28°C; these temperatures can of course occur in most roofs of heated buildings.

1.10.2.5   Proprietary products have been developed to meet the most stringent requirements. Reinforced polyethylene is, typically UV resistant and thermally stable in the temperature range –40°C to +75°C. Its superior strength can also simplify construction because it is much easier to install and less susceptible to damage. These products can also be made fire resistant and/or laminated with metallic foil, to give vastly enhanced vapour resistance.

1.10.2.6 The units for measuring vapour resistance are MN s/g, and these provide the means for comparing different products. Not long ago, it was claimed that a vapour control layer should have a vapour resistance of at least 15 MN s/g; today that idea is totally discredited. Most authorities agree that, even in 'average' conditions the vapour control layer should have a resistance of at least 500 MN s/g. Special products are now available for high humidity conditions.

1.10.2.7 Vapour control layer performance is currently under consideration by European Committee for Standardisation CEN TC 254 WG7.

1.10.3 Installing VCL

1.10.3.1 When installing separate plastic vapour control layers it is essential to ensure all side and end overlaps are sealed to prevent the passage of vapour through the lap. The overlaps should be a minimum 150 mm and sealed with an appropriate highly adhesive double sided sealant tape.

1.10.3.2 Ideally, edge laps should be on top of the liner profiles to allow pressure to be exerted when sealing with the tape. Spacer supports such as ferrules should seal to the vapour control layer and washered screws should be considered for high humidity conditions.

1.10.3.3 As it is impossible to construct a vapour control layer which totally resists every trace of vapour and since it would be undesirable to allow moisture to accumulate in the roof, it is necessary to ensure that any traces of vapour which enter the roof construction can be dispersed. Dispersion of vapour is usually achieved through ventilation. Vapour can be removed through unsealed side laps, provided these are reasonably close together, or through the unfilled ribs in profiled sheets. Ventilation through the ribs depends on air being allowed to enter at the eaves, and escape at the ridge. Ventilated filler blocks may be used to allow this air movement to take place (or the filler blocks may be omitted altogether if insects, vermin, and wind driven rain, are not a threat to the satisfactory occupation of the building). The ventilation is driven by the 'chimney' effect; it therefore becomes more efficient as the roof pitch increases. The efficiency of ventilation through the ribs is also dependant on the overall length of the rib.

It is difficult to quantify the ventilation effects in terms of rib spacing, rib area, roof pitch and slope length, but designers should be aware that short, steep slopes are best in this respect.

1.10.3.4 From the foregoing, it will be appreciated that ventilated ribs are at their least efficient on long shallow pitches. In these cases, it is better to use standing seam systems which have unsealed seams, at relatively close spacing.

1.11 THE BREATHER MEMBRANE

1.11.1 Use of breather membrane

1.11.1.1 A breather membrane is intended to provide additional protection in built-up systems which rely on ventilation through the profile ribs, or any ventilated cavity. Breather membranes have minute pores which permit vapour to flow from one side to the other, but resist the passage of water droplets in the opposite direction. They can be formed by perforating during the manufacturing process, or by bonding non-absorbent fibres in such a way that the pores are created by the intersection of the fibres.

1.11.1.2 The breather membrane should be placed directly over the insulation so that vapour can escape, through its pores, into the ventilated ribs or void. If condensation occurs on the outer sheets, the water cannot wet the insulation. If there is sufficient condensate, and if the roof is steep enough, this condensate will run down the breather membrane to discharge at the eaves; the eaves detail must be designed to permit this drainage. If the pitch is shallow, and if the amount of condensate is small, the water droplets may stand on the breather membrane until climatic conditions improve and it is re-evaporated and ventilated away as a vapour.

1.11.1.3 In order to ensure that the breather fulfils its design purpose, it must be laid according to certain principles. There must be a ventilated space above it, this can be continuous or intermittent (as in the case of profile ribs). It should be reasonably uniformly supported, as sagging sections could promote ponding instead of allowing clear drainage. Rooflights and other penetrations should be designed to ensure that the drainage path is not interrupted. Laps should be a minimum of 150 mm, be arranged to shed condensate down the slope and sealed with an appropriate tape.

1.11.1.4 Just as the vapour control layer should provide the greatest possible vapour resistance, the breather membrane should exert the least possible vapour resistance (compatible with it remaining watertight). There are various opinions as to what is the 'ideal' vapour resistance for a breather membrane. BS 5250 suggests that the typical value is 0.5 MN s/g, and the total range 0. 1 MN s/g to 6 MN s/g.

1.11.1.5 Breather membrane performance is currently under consideration by European Committee for Standardisation CEN TC 254 WG9.

1.11.1.6 Breather membranes must continue to provide their protection throughout the life of the roof. This means that they must be durable, and must not rot or deteriorate through ageing, or under the action of air and water, or changes in temperature. Nor must they be degraded by UV radiation where they discharge into the gutters. Proprietary systems are available which are resistant against tearing, UV degradation, and fire. Some membranes sold as breather layers do not meet all these criteria and therefore the product should be selected on technical performance, and not price, especially when the building will contain a high humidity environment.

1.12 THERMAL INSULATION

1.12.1 Choice of insulation

1.12.1.1 The selection of profiled sheeting cannot be divorced from the subject of thermal insulation because the specifier has the option to use one of the integrated systems.

1.12.1.2 The principal systems requiring a simultaneous commitment on sheeting profile and insulation are:

1. high quality, energy efficient engineered systems and composite panel systems suited to applications involving high humidity;
2. composite and semi-composite panel systems suited to moderately high humidity conditions;
3. bonded sheets (a construction of pre-formed rigid insulation bonded to the underside of profiled sheet) with or without filled profiles. Generally, this construction is used for conditions of low humidity.

1.12.2 Design considerations

1.12.2.1 By improving the standard of thermal insulation beneath the roof sheeting, the outward flow of heat during cold weather is reduced. As it receives less heat from within the building, the temperature of the roof sheeting will fall, thus increasing the risk of condensation

on its underside. Improved insulation, coupled with the increased use of profiled sheet metal roofing and cladding, has added a further dimension to the problem of condensation – the phenomenon known as the 'night sky radiation effect' or 'supercooling' – to which reference is made later.

1.12.2.2 Just as thermal insulation beneath roofing and cladding gives rise to a reduction in surface temperature in winter, exposure to solar heating will result in high sheeting temperatures. This increased range in sheeting temperatures, from maximum to minimum, and the high coefficients of thermal expansion of metals used for sheeting, emphasize the need for adequate provision to cater for thermal movement in the length of profiled metal sheets, particularly in the case of aluminium.

1.12.2.3 Sheet metal cannot absorb and release water in the same way as materials such as fibre reinforced cement. If condensation occurs on the underside of sheet metal roofing in sufficient quantity, it will form water droplets. In a single skin construction this water could drip into the building. Where the cladding is site assembled double skin and the roof has not been designed and built correctly, condensation on the outer sheet could wet the insulation and reduce its thermal efficiency.

1.12.2.4 In the interests of health and safety, consideration should be given to the reaction of thermal insulation when exposed to fire. All plastics-based thermal insulants in common use are combustible and most are capable of generating toxic smoke. Mineral fibres (glass or rock) are non-combustible unless they incorporate flammable bonding agents or coverings.

1.12.2.5 There is a certain amount of heat loss through some roofing and cladding systems where the insulation thickness is reduced locally or where there is a metallic path through the insulation, e.g. at spacers. These are known as thermal bridges and their effects must be taken into account when calculating U-values.

1.12.3 Thermal insulation – physical properties

1.12.3.1 Insulation material provides resistance to the passage of heat energy.

1.12.3.2 The more effective thermal insulants have a low density fibrous or cellular construction and the cells or interstices contain air or some other gas.

1.12.3.3 When selecting an insulant its physical properties need to be considered from various aspects, as the following 'ideal' material properties suggest.

1. An insulant must be unaffected by the most extreme temperature conditions likely to be encountered in storage, in transit and in service.
2. It may require a degree of mechanical strength, rigidity or resistance to tearing, compacting or movement under the influence of its own weight or other superimposed loading.
3. It should not absorb moisture nor suffer deterioration if wetted, whether accidentally during erection or as a result of condensation.
4. It should be non-combustible, non smoke generating and without toxic by-products when subjected to fire.
5. It must discourage infestation and resist fungal growth.
6. It must be compatible with all materials with which it will come into contact, not promote corrosion and be unaffected by accidental contact with solvents or other substances used in its vicinity.
7. It should not constitute a risk to the health and safety of those involved in handling or using the material, or to the occupier of the completed premises.
8. It should have the lowest possible thermal conductivity throughout a life expectancy equal to that of the roof and wall cladding in which it is to be used.
9. Additionally, insulants used in factory made composite panels should form a strong permanent bond with the metal facings, be unaffected by the extremes of temperature and cyclic changes encountered in service, and be able to accommodate any stresses produced by any differential movement of the outer and inner sheets.

1.12.4 Assessment of a system's thermal performance

1.12.4.1 When calculating thermal performance it is important to consider the complete system because each element, including the liner, spacer, insulation and outer sheet, contributes to the total resistance to heat transmission.

1.12.4.2 The term 'element' includes every cavity whether continuous or interrupted (as in many applications involving profiled sheet metal in contact with an insulation layer).

1.12.4.3 The internal and external surfaces exposed to air contributes to the total thermal resistance due to the existence of a boundary layer or surface film for which standard surface resistance values are published.

1.12.4.4 The thermal characteristics of an insulating material is normally expressed as λ, its conductivity value, which is a measure of the rate at which heat will flow through a material when a difference exists between the temperature of its surfaces. It is expressed as W m/m^2 K = W/m K and is independent of thickness. When a thickness of insulating material is known, its thermal resistance, R, can be calculated from

$$R = t/\lambda \ \text{m}^2 \ \text{K/W}$$

Hence for a material (1) having a thickness t_1 and a thermal conductivity λ_1:

$$R_1 = t_1/\lambda_1$$

1.12.4.5 It should be noted that when the moisture content of a material increases, the numerical value of its thermal conductivity increases and hence the thermal resistance decreases. Consequently it is important that thermal insulation materials are kept dry at all times if design performance is to be maintained.

1.12.4.6 Thermal conductivity varies from one material to another and also with density, moisture content and the surface temperatures to which the insulant is exposed in service. Furthermore, certain rigid board insulants, notably foamed polyurethane and polyisocyanurate, undergo a reduction in thermal resistance with time, as the gas retained within the cellular structure is gradually lost by diffusion and replaced by air. This is referred to as 'thermal drift' in the USA, where it has prompted attempts to agree industry-wide design values, allowing for the reduction in thermal efficiency of such insulants. It is advisable therefore, to use published data which take account of these factors, exercising discretion in the choice of value for design purposes.

1.12.4.7 Thermal transmittance is represented by the symbol U in heat transfer formulae and is commonly referred to as the U-value.

1.12.4.8 The U-value is expressed in W/m^2 K, signifying the rate of heat transfer in watts per square metre of surface of the construction for each degree of temperature difference between the design outdoor and indoor air temperatures.

1.12.4.9 Temperature difference is expressed in kelvin (K). A temperature change of one kelvin is numerically equal to a temperature change of one degree Centigrade, thus whichever scale is used the numerical result will be the same.

1.12.4.10 A step-by-step method for determining the U-value for a simple construction is as follows.

1. From published data obtain the value of thermal resistance for each of the elements which make up the total

construction from inside air to outside air, including the inside and outside surface films and any cavity.
2. By simple addition calculate the total thermal resistance.
3. The U-value is found by taking the reciprocal of the value obtained from step (2).

Note: When acquiring data on thermal properties ensure that the units are consistent. S.I. units are used in this publication.

Expressed in mathematical terms:

$$\text{Total thermal resistance, } R$$
$$= R_{si} + R_1 + R_2 + R_3 + R_a + R_{so}$$

where R_{si} = thermal resistance of the inside surface film;
R_1, R_2 and R_3 = thermal resistances of the elements;
R_a = thermal resistance of an air space or cavity;
R_{so} = thermal resistance of the outside surface film.

All thermal resistance values are expressed in m^2 K/W and the thermal transmittance, $U = 1/R$ (W/m^2 K).

1.12.4.11 Example: calculation of thermal transmittance (U-value).

The following example is given to show the principles of the calculation method. It is only applicable to simple constructions where there are no thermal bridges.

Element by element in sequence commencing from outside in:

Element 1: Trapezoidal profiled steel sheet – 0.7 mm thick pre-painted (i.e. non metallic surfaces – hence high emissivity, $R_{so} = 0.04$ m^2 K/W).
Thermal conductivity, $\lambda = 65$ W/m K.

Element 2: Inter-profile cavities.
The thermal resistance of the actual cavity voids can be evaluated in relation to their dimension and their effects proportioned over the roof area, however it is common practice to use the standard resistance as being uniformly effective. $R_{cavity} = 0.09$ m^2 K/W.

Element 3: Mineral wool, to a uniform thickness of 80 mm.
Thermal conductivity, $\lambda = 0.04$ W/m K.

Element 4: Profiled steel liner sheet, 0.5 mm thick pre-painted (i.e. non-metallic surfaces – hence high emissivity, $R_{si} = 0.10$ m^2 K/W).
Thermal conductivity, $\lambda = 50$ W/m K.

Owing to their minimal thickness and high thermal conductivity, the sheet metal roofing and lining contribute very little towards reducing the heat flow through a well insulated construction and are disregarded when calculating the U-value. Surface film effect makes a greater contribution and must be taken into account (see Table 1.2).

Thus, if thermal transmittance, $U = 1/R$ and $R = 2.23$ m^2 K/W, then $U = 1/2.33 = 0.45$ W/m^2 K.

1.12.4.12 The need for a full 80 mm thickness of insulating material is quite apparent and it follows that when spacers are introduced between lining panel and roof sheet the U-value is likely to increase. Tables of

Table 1.2 Calculation of thermal transmittance (U-value)

Element	Thickness, t (m)	Conductivity, λ (W/m K)	Calculation	Thermal resistance, R (m^2 K/W)
External surface film	–	–	Standard value; roof high emissivity normal exposure	$R_{so} = 0.04$
1. Roof sheet	0.0007	65	0.0007/65	$R_1 = 0.00$
2. Profile cavity	–	–	Standard value; roof high emissivity heat flow horiz/upwards	$R_2 = 0.09$
3. Mineral wool	0.080	0.04	0.080/0.04	$R_3 = 2.00$
4. Metal lining panel	0.0005	65	0.0005/65	$R_4 = 0.00$
Internal surface film	–	–	Standard value; roof high emissivity heat flow upwards	$R_5 = 0.10$
Total thermal resistance				$R = 2.23$

insulation thickness required to compensate for thermal bridges at spacers in built-up metal cladding are given in Chapter 3.

1.12.5 Other insulated constructions

1.12.5.1 Site assembled composite systems and factory made composite panels are completely filled with insulation and do not use metal spacer systems. They usually have no significant thermal bridges and their U-values, which take into account the insulation in the profile ribs, are normally quoted by the manufacturer.

1.12.5.2 For advice on the U-values of other special cladding systems such as structural liner trays, contact the manufacturer.

1.12.5.3 No matter which construction is used, it is vital that the insulation material is laid continuously across the whole roof ensuring there are no gaps, particularly at apertures, penetrations, ridge, eaves, etc.

1.13 AIR LEAKAGE

1.13.1 Approved Document L – Air Leakage

Approved Document L has introduced mention of air leakage for the building envelope although no limits have been introduced. Air leakage out of the building is important because it carries away heat, which can only be replaced by additional heating, and the leak may carry moist air into the cladding system with the potential for forming condensation. Similarly air leakage into the building will introduce colder air which may cause interstitial condensation within the cladding. BS 8200 gives some advice on air permeability of wall cladding and the method for calculating heat loss by air leakage. It is therefore important to minimize air leakage through the cladding envelope to reduce energy loss and the risk of condensation.

1.13.1.1 Air leakage can be simply defined as the movement of air in and out of the building which is not for the specific and planned purpose of exhausting stale air or bringing in fresh air. Air leakage cannot be treated, nor can its rate of supply or distribution be controlled in any way.

1.13.1.2 Air leakage should not be confused with water vapour diffusion through the construction. These are two very different envelope performance issues which must be considered during both the design and construction of all cladding systems.

1.13.1.3 Air leakage through the building envelope is driven by three principal mechanisms. These are as follows.

1. The wind – which will produce positive and negative pressures around the building envelope.
2. Mechanical systems – which may produce negative (extract systems) pressures within the building or positive (air supply) pressures within the building.
3. Stack effect – warm air rising and producing a positive pressure effect at high levels within the building envelope.

These driving mechanisms will act on all buildings to some extent at varying times of the year.

1.13.1.4 Air leakage through a building envelope will lead to both ingress of outside air and egress of the internal air. This will lead to either heat loss where heated air within the building is lost through the envelope or heat gain where warmer external air ingresses into cooler temperature controlled buildings. In typical industrial and commercial buildings within the UK it has been estimated that in excess of 50% of the heat loss from the building is attributable to air leakage. In addition to the direct heat losses to which the warm air either egressing or ingressing through the envelope will lead, significant levels of interstitial condensation deposition can occur within multi-component assemblies. Research has shown that up to a hundred times more moisture is moved into an assembly by air leakage than will move into the assembly by vapour diffusion alone (*ASHRAE Fundamentals Handbook,* Chapter 21.4).

1.13.1.5 Air leakage cannot be considered as acceptable ventilation. The level of leakage occurring through the envelope cannot be controlled as it is dependent on the driving mechanisms. Further, the rate of distribution cannot be controlled in any way, nor can the incoming air be treated. It is now widely accepted that for effective ventilation systems to operate within a building, the level of uncontrolled air leakage through the envelope must be restricted. Even in naturally ventilated buildings it is accepted that low levels of envelope air leakage must be achieved if the effective operation of the designed natural ventilation system is to operate efficiently.

1.13.1.6 In most buildings air leakage through the external envelope occurs through the junctions of the various components which form that assembly. This may be junctions between, for instance, masonry to cladding; cladding to glazing or door assemblies; or between components in the cladding assembly itself. Whilst each of these junctions may appear to contribute a very small level of leakage to the building, the overall accumulative effect throughout the envelope can be very significant indeed.

The problems of air leakage are common to both double-skin and composite panel cladding. Like thermal bridges, the areas which need a high standard of workmanship and design consideration are around any junctions to the cladding sheets, these will include:

- ridge, hips and eaves;
- rooflights, smoke vents and pipes;
- corners, sills and parapets;
- windows and doors;
- penetrations, etc.

On many buildings the complicated intersection of metal sheets and flashings are left to the skill of the cladding contractor, and are designed on the job with no guidance on drawings. Although watertight, this on-the-job detailing often results in cavities through the liner and insulation.

1.13.1.7 The designers and installers of cladding systems must pay attention to detail during both the design and installation of the systems. Continuity of the air barrier is critical and should be considered in a similar manner to continuity of the vapour control layer.

1.13.2 Air leakage testing

An increasing number of new build projects now have an air leakage specification which the building must achieve. On completion of the building envelope the building is tested to check compliance with the specification documents. Typically the testing will not concentrate on the specific level of leakage through the cladding system, but will assess the whole building leakage characteristics. A leakage specification will require a building to achieve an envelope leakage rate of x m³/h per m² of external envelope at a pressure differential of 50 Pa. Typical UK standards are 10 m³/h/m² at 50 Pa or 5 m³/h/m² at 50 Pa.

The accepted industry norm for testing complete buildings is a full building fan pressurization test. This effectively involves utilizing a specialist calibrated fan pressurization or de-pressurization rig which supplies or extracts quantifiable levels of air to the building whilst the differential pressure through the building envelope is measured. The results can then be extrapolated to provide a total volume of leakage per square metre of external envelope for that particular building. Further details of this test can be found in BSRIA Technical Note 7/92, *Ventilation Heat Loss in Factories and Warehouses*, or BRE occasional paper, *Determining the Air Tightness of Buildings by the Fan Pressurization Method*, 1987.

1.13.3 Method of test for air permeability of joints in building, BS 6181: 1981

This British Standard paper for testing gives guidelines on apparatus, procedure and test report results. The testing is done under laboratory conditions in a sealed chamber, where means of providing a controlled differential pressure across the test joint is provided.

1.13.4 BRESIM technique – BRE: Information Paper IP 11/90

This is a simplified technique to determine approximately the infiltration and ventilation rates of large and complex buildings. It is an easy-to-use, inexpensive package which involves dispersing a tracer gas within the building. Once the gas has been dispersed and a suitable time elapsed, then the first of two averaged air samples are collected at one or more representative locations. The second sample is taken at the same location after a further, but shorter period of time. Tracer concentrations in samples are subsequently analysed in the laboratory and the hourly air change rate is determined.

1.13.5 References

Additional information can be found in the following papers.

MCRMA Technical Paper No. 10. Profiled Metal Cladding for Roofs and Walls – Guidance Notes on Revised Building Regulation 1995, Parts L and F.
BS 6181: 1981, Air permeability of joints in buildings.
BRE Information Papers; IP11/90; IP 2/93; IP 12/93.

2

PROFILED METAL ROOF AND WALL CLADDING

2.1 INTRODUCTION

2.1.1 Profiled sheet metal

The specifier selecting profiled metal cladding for roof or walls, is faced with a choice from an extensive product range with diverse characteristics. The specifier must consider the following.

2.1.2 Statutory and other requirements

Building Regulations, local requirements and insurance stipulations.
British Standards and Codes of Practice
European Standards ENs and ENVs

2.1.3 Practical considerations

These include:

- weighing the merits of maximum sheet length with fewer end laps against handleability under site weather and location conditions with the attendant risk of damage (see Table 2.1);
- avoiding profiles with features which might affect the ability to ensure secure side laps or their potential for leaks;
- selecting from the range of fixing methods including the 'concealed fix' designs;
- surveying or making an overview of the total roof

system including thermal insulation, practicality of vapour-sealing, lining material and any requirement for a breather membrane;
- making provision for thermal expansion;
- reviewing the system's ease of maintenance and future sheet replacement;
- safety during installation.

2.1.4 Durability

This includes:

- having regard to foreseeable corrosion risks to sheeting, fasteners, spacers, etc.;
- considering the lifespan of material (base metal and coatings on both faces) within extremes of temperature and thermal movement;
- predicting the period to first major maintenance;
- estimating the degree of colour change resulting from weathering and UV radiation;
- ensuring compatibility with and sufficiency of the lifespan of other materials.

2.1.5 Appearance

This includes:

- colour, texture, effect of profile on reflection and shadow;

• any facility for introducing curvature and other features of architectural interest.

2.1.6 Cost

This includes the obvious elements, but also anticipating labour-intensive details such as double lap-sealing, taping to stop bimetallic contact or to provide a thermal break; and the consequential cost of entrusting specialist's work to inexperienced contractors.

Note: The lowest price may be the most costly in the long run! This prophecy will be as applicable to the roofing contractor as to the building owner.

2.1.7 Types of metal substrate

The most popular base materials are steel and aluminium alloy with stainless steel available for service in abnormal environments or where very long life is a requirement.

2.1.8 Steel substrate

A steel substrate should have a primary corrosion resistance, for example by hot-dip coating with zinc, before any subsequent treatment by a factory applied organic coating such as plastisol or PVF_2.

BS EN 508–1 (scheduled for implementation in 1999) will be the British Standard for products covering self-supporting profiled metal, with BS EN 508–3 covering stainless steel. The standard includes material specifications, profile tolerances and recommendations for ordering, delivery and storage.

In the UK the normal minimum recommended specification for hot-dip zinc coating is 275 g/m^2 on steel coil of grade S 220 GD + Z275 (steel number 1.0241 + Z275) to BS EN 10147.

Alternative metallic coatings incorporating small quantities of aluminium are increasingly being manufac-

tured to BS EN 10214. The equivalent material strength specification to that given for hot-dip zinc would be, hot-dip coated with 255 g/m^2 on steel coil of grade S 220 GD + ZA255 (steel number 1.2041 + ZA255) to BS EN 10214. There are a number of different trade names for this type of coating. The steel producers claim that the addition of small quantities of aluminium improves the durability of the zinc coating.

2.1.9 Aluminium substrate

Aluminium may either be used in natural mill finish or given a chemical treatment (often referred to as conversion coating) to modify the initial appearance of the natural surface, or to give a key for a decorative finish.

BS EN 508–2 will be the corresponding standard for self-supporting profiled aluminium. The normal grades of aluminium for profiling are EN AW 3004 and EN AW 3105 to BS EN 485.

2.2 DURABILITY

2.2.1 Life expectancy of profiled metal

The life expectancy of profiled metal is quoted in various ways and the specifier of the sheeting should be careful only to compare durability figures on a like for like basis.

2.2.2 Period to repaint decision

Steel companies supplying plastisol or other organic coated sheet tend to define colour life by suggesting a *period to repaint decision*. This is the time when the building owner must decide whether to re-paint. Repainting will extend the life of the sheeting. The integrity of the sheeting may be endangered without repair.

Plastisol coated steel is more durable in the pale colours which are usually recommended for roofing. The durability periods quoted for other colours vary widely between manufacturers and should be checked before purchase especially if intended for roofing.

2.2.3 Life to first maintenance

Some companies continue to quote a life to first maintenance; this can imply to the building owner that no maintenance is necessary on the roof during this period. The phrase 'life to first maintenance', which only applies to the sheet coating, encourages serious neglect of the roof envelope and gutters. All roofs need regular inspection and maintenance; see Chapter 10 for recommendations.

Table 2.1 Guide to profiled sheet weights

Cladding system	Typical design weight (kg/m^2)		
	Aluminium	Coated steel	Stainless steel
Single skin	5	7	7
Built-up with metal liner and mineral insulation quilt 0.45 U-value	11	13	13
Factory insulated foam composite panel	8	10	N.A.

Note: A typical 35 mm deep trapezoidal profiled metal outer sheet and 18 mm deep profiled liner sheet have been assumed for this table.

2.2.4 Functional life

Aluminium suppliers often quote a functional life for the bare aluminium sheet but it may lose any aesthetic colour coating before this period expires.

2.2.5 Selection by the specifier

The specifier of the profiled sheeting is responsible for selecting the material type and supplier from a wide range of quoted, and in some instances, guaranteed durability periods.

2.2.5.1 The most popular coating for the external facing of steel roof panels is embossed PVC plastisol in a thickness of 200 μm. This coating may also be used for vertical cladding, alternatively smooth finish PVF_2 may be used for wall panels because of its higher gloss level and better resistance to chalking or fading. PVF_2 is not recommended for roofing purposes, as the paint finish is less resistant to damage and has a shorter life expectancy. Polyester and silicone polyester, are the usual coatings for steel lining sheets. Profiled steel roofing sheets are typically 0.7 mm thick with 0.5 mm often used on walls. Profiled steel facings on factory insulated composite panels are typically 0.5 mm thick with liners for built-up and composites from 0.4 mm thick.

2.2.5.2 Aluminium profiled sheeting is sometimes preferred in special applications; these could include very humid conditions, or a severe marine environment. External aluminium profiled sheeting is typically 0.9 mm in thickness, and linings 0.4 mm or 0.5 mm. Aluminium profiled factory insulated composites are typically 0.7 mm thick on the exposed face.

2.2.5.3 Aluminium does not usually require coating for its protection, so mill finish preferably with embossed surface may be used. Where coloured finishes are required on aesthetic grounds, PVF_2 and abrasion resistant polyurethane are the popular choices for the external skin, while polyester is usually specified for linings.

Selection by the specifier of the lowest initial cost sheeting may not be the optimum choice for the building owner.

2.3 APPLICATIONS

2.3.1 Choice of profile

2.3.1.1 The greatest choice is of profile, including symmetric and asymmetric trapezoidal form, with or without an array of stiffening ribs and in a range of profile depths from 19 mm to about 60 mm at pitches from 75 mm to over 330 mm.

2.3.1.2 Concealed-fix and standing-seam systems extend the profile pitch and depth to around 600 mm and 75 mm respectively.

2.3.1.3 The range of roofing profiles is far exceeded by those for use as vertical cladding, where appearance can be enhanced by the choice which exists to run the profile vertically, horizontally, or diagonally. Diagonal cladding requires extensive site cutting of profiled sheet and wastage which should be considered at the design stage.

2.3.1.4 Normal material thickness for external use ranges from 0.5 mm to 1.2 mm for walls and 0.7 mm upwards for roofs.

2.3.1.5 Wall panels, especially flat (cassette) panels with bonded insulation, are available with factory fitted gaskets in their joints. These systems are outside the scope of profiled sheeting described in this text. Details should be obtained direct from the specialist manufacturer.

2.3.1.6 Reference to section 2.2, Durability, shows that the choice of a surface finish involves much more than selecting a colour.

2.3.1.7 It is common practice for roll-forming of the profile to be carried out on the pre-coated strip material. Thus the substrate and surface coatings are subjected to bend stresses. It is important therefore, that the durability of the finish is not impaired either by crazing of the coating or by weakening of the bond between substrate and coating. Neither must the thickness of coating seriously be reduced on external radii.

2.3.1.8 The finish should be resistant to scratching and abrasion (accidental or otherwise) and capable of surface repair. It should be colour-fast to prevent any subsequent replacement sheet being too conspicuous. The selected finish should withstand the appropriate environmental conditions (e.g. industrial, agricultural, marine, etc.) and any high surface temperatures resulting from solar radiation.

2.3.2 Use of light colours

2.3.2.1 The use of light colours has a significant effect on surface temperature in full summer sunlight. Reductions of around 40°C, below the temperature that a black surface would reach, are attainable. White finishes should be used with caution, for any paint technologist will confirm that it is the most difficult colour to use for accurate colour matching.

Important note: Where organic coatings subject to shade variation such as white are used, the same production batch should be used for all the profiled sheeting, flashings and trims, on that contract. Metallic coatings, such as PVF$_2$ silver, may exhibit directional shade variation. These coatings should therefore be from the same production batch and be laid in the same direction.

2.3.2.2   All reputable manufacturers will provide full data on their surface finishes assuring compliance with relevant Standards and substantiated by test reports.

2.4   CONTROL OF CONDENSATION

The risk of condensation has been considered in Chapter 1. All metal roofs whether of hybrid or composite construction require measures to limit the movement of moisture laden air into the insulated structure.

2.4.1   Hybrid construction

The experience of the metal cladding industry over the last 10 years, with roofs of 4° to 10° pitch, is that a breather membrane is not required in some circumstances. Any requirement depends on the building's internal environment and the application. In normal applications, provided the vapour control layer has been properly installed, the amount of condensation is likely to be small and no breather membrane is required.

2.4.2   Internal humidity grades

The normal applications for built-up metal roofing are defined in the Metal Cladding and Roofing Manufacturers Association Technical Paper No. 10, as Grade A.

2.4.2.1   Grade A – Normal humidity
This applies to factories and warehouses for normal manufacturing and storage purposes where the occupants or processes do not add significant quantities of water vapour to the atmosphere.

2.4.2.2   Grade B – Medium humidity
This applies to buildings where large numbers of people congregate, for example public meeting halls, supermarkets and offices and buildings such as sports halls or where heating is intermittent, e.g. church halls.

2.4.2.3   Grade C – High humidity or special environments
This applies to swimming pools and buildings containing liquids stored in open containers or where water is used in manufacturing, cleaning or storage processes, e.g. ice rinks, cold and chill stores, etc.

2.4.2.4   The appropriate control measures for moisture vapour related to the above internal environments is given in Chapter 3 for insulated metal sheeting and Chapter 4 for factory insulated composite panels.

2.5   AIR LEAKAGE THROUGH THE BUILDING ENVELOPE

2.5.1   Movement of air

2.5.1.1   Most solid materials permit the movement of air to some extent, especially at jointing intersections. Whenever there is a difference in air pressure either side of that joint, then a movement of air can take place.

2.5.1.2   The amount of air leakage through the building envelope will depend upon the internal heat build-up from heating systems and mechanical processes, causing the stack effect (warm air rising). Externally, the wind pressure will also be a governing factor. The resultant leakage can substantially affect the heating cost of that building. However, it would not be sufficient enough to simply increase the U-value for the building envelope. Up to 50% of the heating energy required to heat a large industrial building may be required to compensate for air leakage losses through joints, junctions, etc. Although adequate ventilation is essential for the health, safety and comfort of the building occupants, excessive ventilation may also lead to energy waste and discomfort.

2.5.1.3   In the roofing and cladding industry, air leakage may be reduced by good design and site practice for both built-up insulated and composite panel systems.

2.5.1.4   The following list details some of the design considerations which are often left to the experience of the roofing contractor. The effectiveness of this detailing relies almost 100% on site workmanship, and failure to achieve high standards on site may result in excessive heat loss and possibly interstitial condensation.

2.5.1.5   The metal cladding itself has a very good resistance to air, however to provide a continuous barrier, the detailing of the following intersections to both the roof and walls must be considered.

Side and end laps

- Built-up insulated. The metal liner sheets may be sealed at the edges with a butyl mastic strip or a self-adhesive tape. Tape is preferable for adjustable side lap joints; this is fixed over the joint and facilitates inspection from above before the insulation and roof sheet are installed.

- Composite panels. At end lap positions, composite panels may be bedded down at both ends on to the supporting structure. At side laps, the panels can be sealed with sealant where the inner skins join.

Ridge, hips and valleys

- Built-up insulated. The flutes of the profile metal liner may be closed with sealed profiled foam fillers, which can be raked cut to suit hip and valley details.
- Composite panels. The panel may be bedded down on a strip sealant, fixed to an internal closure flashing.

Roof penetrations

Sealing around the liner intersection with roof lights, pipe penetrations and ventilators may be achieved by careful positioning of seals, flute fillers or pipe flashings.

Parapets, sills and corners

These are often details which are totally neglected, however they are one of the most likely causes of draughts or air leakage. Sealing flashing junctions to sheeting/panels with sealants and blocking corrugations with profiled polyethylene flute fillers are usual ways of dealing with these areas.

Windows and doors

Where the liner cladding intersects windows and doors, etc., the joints should be sealed. The designer should ensure that all opening elements of the door or window have sealant strip draught excluders. Where doors are continuously being opened, it may be more feasible to include a draught lobby, automatic door or air curtain, or for loading docks, polythene strip curtains and surrounds.

2.5.1.6 Where ventilation louvres are required to discharge fumes or allow air intake for cooling processes they should be specifically designed for their intended purposes and possible allowance for complete closure when not in use.

2.5.1.7 For buildings with a high internal humidity environment, consideration for the sealing of the fixings must be given. Fixings should have sealing washers, and spacer supports, where used, should have sealing lips or be bedded on sealant.

2.5.1.8 For insulated systems, a separate vapour control layer may be considered as an aid to reducing air leakage, provided that all the penetrations are fully sealed.

2.5.1.9 On projects today, the ability of the cladding system to resist the passage of air may be tested to check compliance with the specification documents prior to completion. There are technical papers available which give details for testing of air leakage through joints, see Chapter 1.

2.6 ROOF PITCH AND SHEETING LAPS

2.6.1 Weathertight envelope

2.6.1.1 The primary function of profiled metal roofing and cladding is to create a weathertight envelope. This involves fixing the sheets to overlap at sides and ends to resist the entry of rain water and snow under all predictable weather conditions.

2.6.1.2 With contemporary design concepts calling for ever lower roof pitches, sealants within the laps must be relied upon to prevent the ingress of water. Alternatively, below 4° concealed fix sheeting systems should be considered.

2.6.1.3 Problems can arise where inadequate allowance has been made at the building design stage for:

1. general construction tolerances;
2. deflection of sheeting, which if excessive can allow ponding;
3. deflection of structure which may reduce design pitch by at least 1°;
4. end-laps, which should be avoided below 4° finished pitch, i.e. 5° design pitch.

2.6.1.4 The roof profile must be capable of carrying storm water to BS 6367 within the capacity of the sheeting profile without over-topping the side laps.

2.6.1.5 The recommended minimum finished pitch for profiled roofing is 4° or a gradient of 1 in 14; designers should consider concealed fix systems, increasing the roof pitch or profile depth for roofs below a design pitch of 5°.

2.6.1.6 Leakage is mainly the result of the dynamic effects when water is carried through the laps by the force of the wind or capillary action due to the difference in pressure acting across the lap. Therefore it is essential that lines of sealant are continuous.

2.6.1.7 Many variables affect the performance of laps, therefore selection of profile design is important and should not be decided on the basis of lowest cost. Among the variables to be considered are:

- topography;
- building height and proportion;
- rate of water flow over the roof;
- coincident wind direction and speed;
- effect of local gusting and turbulence set up around profiles;
- pressure differentials across the sheeting.

2.6.2 Side and end laps

The minimum end lap should normally be 150 mm for steel and aluminium profiled roofing, with the proviso

that an increase to 200 mm is justified at moveable end laps in aluminium sheeting when proprietary expansion-joint components are used (see Figures 2.1 & 2.2).

For pitches down to 1.5° a secret fix or standing seam profile should be used incorporating, if necessary, the manufacturer's recommended end lap details.

The roof pitches quoted above are equivalent to the following gradients:

5° = 1 in 12, 7° = 1 in 8, 15° = 1 in 3.8, 35° = 1 in 1.6

Notes relating to Table 2.2

1. The positioning of fasteners in profiled sheeting is to be in accordance with details provided by the profile manufacturer.

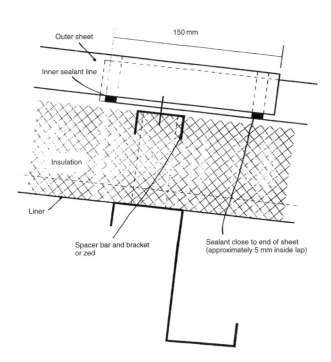

Figure 2.1 Built up insulated roof – end lap.

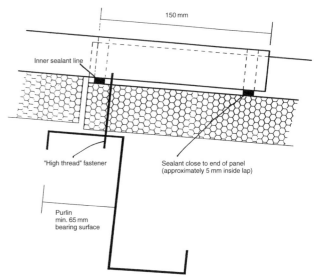

Figure 2.2 Through fixed composite roof panel – end lap.

Table 2.2 End laps and side laps

Roof cladding

Pitch	Min. end lap (mm)			Max. side lap fastener spacing (mm)		
	Profiled steel	Profiled aluminium	Profiled rooflights	Profiled steel	Profiled aluminium	Profiled rooflights
5°–7°	150 (D)	150(DS)	150(DS)	450 (S)	450 (S)	300 (S)
7°–35°	150 (D)	150(DS)	150(DS)	450 (S)	450 (S)	300 (S)
Over 35°	150 (S)	150 (S)	150 (S)	450 (S)	450 (S)	300 (S)

Wall cladding

Vertical profile	100 (U)	100 (U)	100 (S)	500 (U)	450 (U)	300 (S)
Horizontal laid profile	150 (S)	150 (S)	$	450 (U)	450 (U)	$

(D)	Double continuous row of sealant as per Figure 2.1, i.e. British Steel's recommendations.
(DS)	Double continuous row of sealant.
(S)	Single continuous row of sealant.
(U)	Unsealed.
$	May require special treatment, contact profile manufacturer for advice.

2. End lap sealing for profiled steel should be in accordance with British Steel's current recommendations unless the profile manufacturer advises otherwise. Fasteners at end laps are to be in accordance with the profile manufacturer's recommendations. The end lap detailing apply equally to sheeting and related components, e.g. traditional soakers and flashings.

3. The profile manufacturer's recommendations must be followed when fixing concealed-fix and composite systems, as these products have special requirements.

4. In roofing applications, side lap fasteners should be fixed along the centre line of the profile crown, with the seal on the weather side.

5. It is recommended that all side laps in profiled metal roof sheeting be of a supported type, i.e. with a full under roll, particularly if a shallow profile (below 30 mm depth) is used.

6. Specifiers should be warned of the increased risk of leakage through side laps when using shallow roofing profiles (particularly on long roof slopes) regardless of pitch. A deeper (greater than 30 mm) profile is always preferable for roof pitches of 15° and below.

7. Use of sealants with strong adhesive properties should not be considered as a substitute for side lap fasteners

8. End laps in plastic rooflights are to be fixed in each profile and the largest sealed washers that can be accommodated in the profile valley, up to 32 mm diameter, used to spread the fixing load. Wide troughs may require two fasteners in each.

9. Side lap fasteners in vertical cladding may be fixed either through the crown or web of the profile, or may indeed be unnecessary when a full profile overlap is provided, but the objective must in any case be to resist the entry of water in storm conditions whilst achieving a neat appearance.

10. Because of the potential for a considerable length of wall cladding with the profile running horizontally, extra attention must be given to the incorporation of movement joints. This is particularly important when aluminium is used because of its high coefficient of linear expansion.

2.7 METAL FLASHINGS

2.7.1 Design

2.7.1.1 On many buildings it is the flashings that are noticed, particularly on curved roofs, and yet the vital junctions and fixings are often left to the experience of the fixer on site. Complaints to roofing contractors are often caused by leaks and appearance of flashings. Wind damage to buildings with metal cladding is frequently limited to flashings which should have been more securely fixed. The perimeter of most building designs is the location of the maximum wind pressures and the flashings.

2.7.1.2 Flashings are usually fabricated from flat metal sheet and supplied to order and design. They should be designed to perform their function as weatherproofing components with due regard to the way in which they can affect appearance.

2.7.2 Recommendations for the design of metal flashings

1. Minimum thickness of metal

External use	Steel	0.7 mm
	Aluminium	0.9 mm
Internal use	Steel	0.4 mm
	Aluminium	0.7 mm

2. Avoid expanses of flat metal in excess of 200 mm by introducing bends, folds, swaging or similar measures to impart stiffening or provide additional support.

3. Use the same material coating specification as the cladding material, use prime material and ask for proof of source. Guarantees for durability of the precoated material are available for flashings as well as the profiled sheets.

4. Avoid using dissimilar materials and coatings to the cladding materials.

5. Allow for joints at 3 m normal lengths, with 6 m maximum available from some suppliers for the simpler flashing shapes.

6. Longitudinal edges should be returned to form a welt or, alternatively, a stiffening bend may be incorporated within 50 mm of the edge.

7. Butt straps used at flashings should be not less than 150 mm in length and used with sealant (see Figure 2.3).

8. Where overlaps occur in flashings a lap of 150 mm is recommended on roofs and 100 mm on walls.

9. Flashings should be secured to the sheeting using appropriate secondary fasteners at 450 mm maximum centres. The fastener should be at a minimum of 12 mm from the edge of the flashing. Lap joints should be fixed at 75 to 100 mm centres.

10. Sealant should be positioned on the weatherside of fasteners where a single line is used. Sealant materials should be good quality as used for the main cladding areas, see Chapter 8.

11. Provision should be made for thermal expansion in all aluminium flashings. If a butt strap is used at an

expansion joint it should be fixed to one flashing only to allow for movement.

12. Prefabricated corner flashings can be supplied in preference to junctions cut and folded on site. Where used, the correct fixing sequence should be made clear to the fixers.

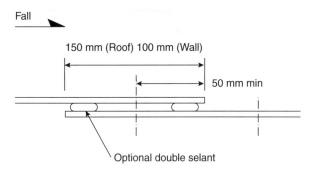

(a) Lap joint, single or double sealant strip (steel)

(b) Steel butt strap, single wide sealant strip each side

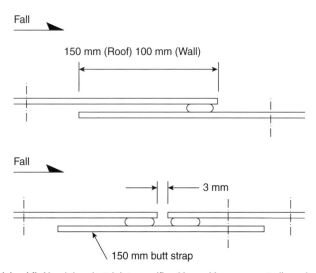

(c) + (d) Aluminium butt joint or unfixed lap, with movement allowed at each joint to avoid thermal movement accumulating in a long length. Low modulus silicone sealant is one suitable sealant. Fasteners should be aluminium sealed rivets or stainless steel screws.

Figure 2.3 Lap joints – flashing (reproduced with permission of the MCRMA).

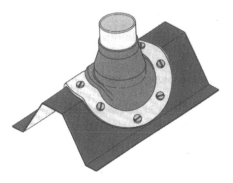

Figure 2.4 Adaptable pipe flashing.

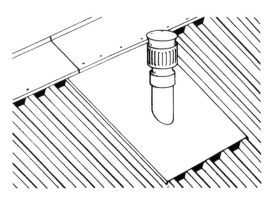

Figure 2.5 Opening through apron flashing to ridge.

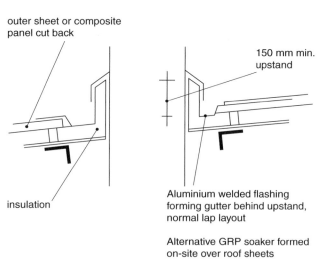

Figure 2.6 Typical large penetration detail.

2.7.3 Roof penetrations

2.7.3.1 Flashings around penetrations for pipes and vents, etc. should be detailed at the design stage as it is essential that vapour control and breather membranes are continuous around the openings.

2.7.3.2 With aluminium cladding, large aluminium soakers can be fabricated by on-site welding. Special care needs to be taken for fire safety and to allow free drainage around the soaker. On-site welding is a specialist operation and the advice of a welding expert should be sought and heeded (see Figure 2.6).

2.7.3.3 On-site GRP flashings can be made around penetrations in a similar manner to on-site welding for aluminium. This is also a specialist operation on which the advice of an expert applicator should be sought and heeded.

2.7.3.4 Small penetrations such as vent pipes can be weathered with pre-formed EPDM or silicone rubber bellows which are supplied with sealing instructions. On roofs, water must be able to drain past the penetration, which should not block the profiled sheet trough. Where blockage would occur it is necessary to fit a flat apron flashing back to the ridge or use an alternative type of soaker (see Figure 2.4 and 2.5).

3

BUILT-UP METAL ROOFING SYSTEMS

3.1 INTRODUCTION

In this chapter guidance is given on built-up metal sheeting which is referred to as hybrid construction in BS 5427. Factory insulated composite panels are dealt with in Chapter 4, site assembled composites may in general be considered as built-up systems.

Secret-fix profiles and curves are included in this chapter, although insulated composite systems are available.

The performance of the profiled metal and its durability has been considered in the previous chapters.

3.2 ASSESSMENT OF THE U-VALUE FOR A CLADDING SYSTEM

3.2.1 Built-up systems with spacers

The general principles employed to select insulation materials and calculate U-values are given in Chapter 1.

3.2.1.1 The U-value or thermal transmittance value required by the Building Regulations for industrial and commercial roofs is 0.45 W/m² K but the regulations require that the U- value calculation must now take any repeating thermal bridges into account. In site assembled double skin constructions this means that the performance of the spacer system must be considered. If the liner profile compresses the insulation locally within the cavity it will also create a thermal bridge and should be included in the U-value calculation.

3.2.1.2 Spacers are used at every purlin and/or rail position, so they repeat at typically 1.8 m centres over the entire roof or wall. Although designed with a thermal break, the spacer system includes a steel section which can reduce the depth of insulation and create a thermal bridge.

3.2.1.3 Approved Document L; 1995 states that the thermal bridging effects should generally be taken into account using the proportional area calculation method. However the Department of the Environment, Transport and the Regions (DETR) (which is responsible for the Approved Document and the Building Regulations), now says that where there are metal bridges through the insulation layer, this proportional area calculation method can lead to a significant underestimate of heat losses. The more accurate finite element analysis method should be used.

3.2.1.4 The following tables which were prepared by BRE and published by DOE (now DETR) show the insulation thickness required to achieve a U-value of 0.45 W/m² K using zed spacer systems. These values were calculated using finite element analysis. The values assume there is no compression of the insulation by the liner profile. The arrangement of insulation at the spacer and the depth of ferrule can make a significant difference to the thickness of insulation required. The tables also show the effects of using different purlin centres and the following insulation materials:

typical fibreglass insulation   0.040 W/m K thermal conductivity

typical rock fibre insulation   0.037 W/m K thermal conductivity

low conductivity fibreglass   0.035 W/m K thermal conductivity

Rail and bracket spacer system

3.2.1.5   Rail and bracket spacers can reduce thermal bridging and, provided the insulation is laid between the rail and liner and brackets are at 1 m centres minimum, the DOE allow the thermal bridging effect of the rail to be ignored in U-value calculations. The insulation thickness required to achieve a U-value of 0.45 W/m² K with a typical rail and bracket is shown in Table 3.1.

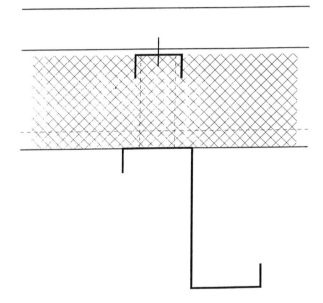

Figure 3.1 Rail and bracket spacer system.

Table 3.1 Insulation thickness for a typical rail and bracket spacer

| Purlin spacing (m) | Required insulation thickness (mm) | | |
| | Thermal conductivity of insulation (W/m K) | | |
	0.035	0.037	0.040
1.0–3.0	67*	77	83

* For insulation thicknesses of less than 80 mm there is an air gap between the top of the insulation and the underside of the outer sheet so that the total thickness of insulation and air gap equals 80 mm.

No insulation under zed spacer

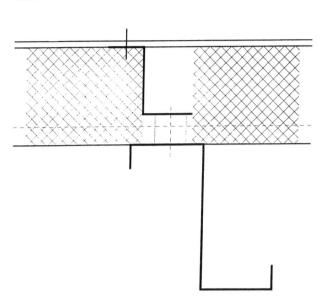

Figure 3.2 No insulation under zed spacer.

Table 3.2 Insulation thickness for no insulation under zed spacer

| Purlin spacing (m) | Required insulation thickness (mm) | | |
| | Thermal conductivity of insulation (W/m K) | | |
	0.035	0.037	0.040
1.0	109	116	123
1.8	90	96	102
3.0	83	89	94

These values also apply to fire rated wall designs where the system has been tested with insulation laid over the zed spacer and compressed locally by the outer sheet.

25 mm insulation tucked under zed spacer

Table 3.3 Insulation thickness for 25 mm of insulation tucked under zed spacer

| Purlin spacing (m) | Required insulation thickness (mm) | | |
| | Thermal conductivity of insulation (W/m K) | | |
	0.035	0.037	0.040
1.0	83	90	96
1.8	73*	84	91
3.0	71*	82	88

* For insulation thicknesses of less than 80 mm there is an air gap between the top of the insulation and the underside of the outer sheet so that the total thickness of insulation and air gap equals 80 mm.

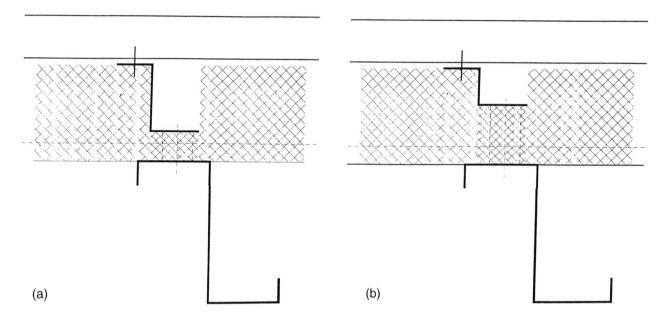

(a)   (b)

Figure 3.3 Insulation tucked under zed spacer: (a) with 25 mm ferrule; (b) with 50 mm ferrule.

50 mm insulation tucked under zed spacer

Table 3.4 Insulation thickness for 50 mm of insulation tucked under zed spacer

| Purlin spacing (m) | Required insulation thickness (mm) | | |
| | Thermal conductivity of insulation (W/m K) | | |
	0.035	0.040	0.040
1.0	69*	81	87
1.8	68*	80	86
3.0	67*	79	85

* See footnote to Table 3.3.

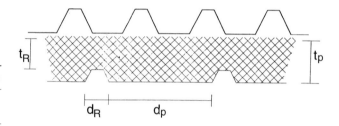

Figure 3.4 Reduction of insulation thickness when compressed by liner profile.

3.2.1.6 The values shown in the tables are the theoretical minimum insulation thicknesses required, ignoring any effect of the liner profile. If the liner ribs compress the insulation a greater thickness will be required. If the liner profile depth is less than 25 mm and the pitch is at least 250 mm the increase in thickness can be determined using the following formula (see Figure 3.4):

$$\Delta t = t_p - \frac{1}{\left[\dfrac{d_R}{(d_R + d_p) \times t_R} + \dfrac{d_p}{(d_R + d_p) \times t_p} \right]}$$

3.2.1.7 The effect of insulation compression by other liner profiles should be determined by finite element analysis.

3.2.1.8 If the insulation is going to be arranged in any other way at the spacer, the correct thickness to use must be established by finite element computer analysis.

3.2.1.9 In practice the mineral fibre insulation quilt used is normally selected from a range of standard thicknesses, such as 83 mm, 100 mm etc., and a thickness greater than or equal to the required thickness should be used.

3.3 VAPOUR CONTROL MEASURES

3.3.1 Vapour control with insulated built-up metal sheeting

3.3.1.1 With typical built-up metal cladding, i.e. 'hybrid construction' (as defined in BS 5427), the liner panels with all side and end laps sealed and washered fasteners used to fix the liner panel may be classed as the

VCL, except for high internal humidity conditions, see below. The seals may be mastic sealants within the laps, or adhesive tapes fixed over the laps from above. Both methods are dependant on the standard of workmanship and the strength and condition of the liner edge. Seals which are omitted, or fail to adhere, can negate the whole vapour control method. Tapes offer better prospects for inspection during construction, but may be more vulnerable later as the building experiences movements due to wind, snow and changes of temperature. Any fasteners which penetrate the liner should have compressible washers to maintain the vapour seal.

3.3.1.2   In the past, ordinary 500 gauge polyethylene sheeting has sometimes been used as a vapour control layer. The tear and vapour resistance of this sheeting is so low that it really should not be considered as a viable option when designing or choosing vapour control systems.

3.3.1.3   The use of thicker gauges (e.g. reinforced 1000 gauge or more) can be a help, but the material must be 'virgin', because recycled products can have minute pores or pinholes which drastically reduce their vapour resistance. Prospective users should be aware that ordinary polyethylene may not be an appropriate material for use as a vapour control layer in some constructions, as it can become thermally unstable at temperatures as low as 28°C; these temperatures can of course occur in most roofs of heated buildings.

3.3.1.4   Proprietary products have been developed to meet the most stringent requirements. Reinforced polyethylene is, typically UV resistant and thermally stable in the temperature range –40°C to +75°C. Its superior strength can also simplify construction because it is much easier to install and less susceptible to damage. These products can also be made fire resistant and/or laminated with metallic foil, to give vastly enhanced vapour resistance.

3.3.1.5   The units for measuring vapour resistance are MN s/g, and these provide the means for comparing different products. Not long ago, it was claimed that a vapour control layer should have a vapour resistance of at least 15 MN s/g; today that idea is totally discredited. Most authorities agree that, even in 'average' conditions the vapour control layer should have a resistance of at least 500 MN s/g. Special products are now available for high humidity conditions.

3.3.1.6   Vapour control layer performance is currently under consideration by European Committee for Standardisation CEN TC 254 WG7.

3.3.2   Installing VCL

3.3.2.1   When installing separate plastic vapour control layers it is essential to ensure all side and end overlaps are sealed to prevent the passage of vapour through the lap. The overlaps should be a minimum 150 mm and sealed with an appropriate highly adhesive double sided sealant tape.

3.3.2.2   Ideally, edge laps should be on top of the liner profiles to allow pressure to be exerted when sealing with the tape. Spacer supports such as ferrules should seal to the vapour control layer and washered screws should be considered for high humidity conditions.

3.3.2.3   As it is impossible to construct a vapour control layer which totally resists every trace of vapour and since it would be undesirable to allow moisture to accumulate in the roof, it is necessary to ensure that any traces of vapour which enter the roof construction can be dispersed. Dispersion of vapour is usually achieved through ventilation. Vapour can be removed through unsealed side laps, provided these are reasonably close together, or through the unfilled ribs in profiled sheets. Ventilation through the ribs depends on air being allowed to enter at the eaves, and escape at the ridge. Ventilated filler blocks may be used to allow this air movement to take place (or the filler blocks may be omitted altogether if insects, vermin, and wind driven rain are not a threat to the satisfactory occupation of the building). The ventilation is driven by the 'chimney' effect; it therefore becomes more efficient as the roof pitch increases. The efficiency of ventilation through the ribs is also dependant on the overall length of the rib. It is difficult to quantify the ventilation effects in terms of rib spacing, rib area, roof pitch and slope length, but designers should be aware that short, steep slopes are best in this respect.

3.3.2.4   From the foregoing, it will be appreciated that ventilated ribs are at their least efficient on long shallow pitches. In these cases, it is better to use standing seam systems which have unsealed seams, at relatively close spacing.

3.3.3   Breather membrane

3.3.3.1   A breather membrane is intended to provide additional protection in built-up systems which rely on ventilation through the profile ribs, or any ventilated cavity. Breather membranes have minute pores which permit vapour to flow from one side to the other, but resist the passage of water droplets in the opposite direction. They can be formed by perforating during the manufacturing process, or by bonding non-absorbent fibres in such a way that the pores are created by the intersection of the fibres.

3.3.3.2   The breather membrane should be placed directly over the insulation so that vapour can escape, through its pores, into the ventilated ribs or void. If condensation occurs on the outer sheets, the water cannot wet the insulation. If there is sufficient condensate, and if the roof is steep enough, this condensate will run down the breather membrane to discharge at the eaves; the eaves detail must be designed to permit this drainage. If the pitch is shallow, and if the amount of condensate is small, the water droplets may stand on the breather membrane until climatic conditions improve and it is re-evaporated and ventilated away as a vapour.

3.3.3.3   In order to ensure that the breather fulfils its design purpose, it must be laid according to certain principles. There must be a ventilated space above it, this can be continuous or intermittent (as in the case of profile ribs). It should be reasonably uniformly supported, as sagging sections could promote ponding instead of allowing clear drainage. Rooflights and other penetrations should be designed to ensure that the drainage path is not interrupted. Laps should be a minimum of 150 mm, be arranged to shed condensate down the slope and sealed with an appropriate tape.

3.3.3.4   Just as the vapour control layer should provide the greatest possible vapour resistance, the breather membrane should exert the least possible vapour resistance (compatible with it remaining watertight). There are various opinions as to what is the 'ideal' vapour resistance for a breather membrane. BS 5250 suggests that the typical value is 0.5 MN s/g, and the total range 0. 1 MN s/g to 6 MN s/g.

3.3.3.5   Breather membrane performance is currently under consideration by European Committee for Standardisation CEN TC 254 WG9.

3.3.3.6   Breather membranes must continue to provide their protection throughout the life of the roof. This means that they must be durable, and must not rot or deteriorate through ageing, or under the action of air and water, or changes in temperature. Nor must they be degraded by UV radiation where they discharge into the gutters. Proprietary systems are available which are resistant against tearing, UV degradation, and fire. Some membranes sold as breather layers do not meet all these criteria and therefore the product should be selected on technical performance, and not price, especially when the building will contain a high humidity environment.

3.3.4   Hybrid construction

3.3.4.1   The experience of the metal cladding industry over the last 10 years, with roofs of 4° to 10° pitch, is that a breather membrane is not required in some circumstances. Any requirement depends on the building's internal environment and the application. In normal applications, provided the vapour control layer has been properly installed, the amount of condensation is likely to be small and no breather membrane is required.

3.3.4.2 Internal humidity grades

The normal applications for built-up metal roofing are defined in the Metal Cladding and Roofing Manufacturers Association Technical Paper No. 10, as Grade A, which is defined in Chapter 2 and reiterated here for reference:

Grade A – Normal humidity

Factories and warehouses for normal manufacturing and storage purposes where the occupants or processes do not add significant quantities of water vapour to the atmosphere.

3.3.4.3   Breather membranes should be considered for higher humidity conditions such as Grade B use.

Grade B – Medium humidity

Buildings where large numbers of people congregate, for example public meeting halls, supermarkets and offices. Buildings such as sports halls or where heating is intermittent, e.g. church halls.

3.3.4.4   A breather membrane should be included when the building has a high humidity internal environment (Grade C) and then it should only be used in conjunction with an effective vapour control layer, eaves to ridge ventilation through the profile ribs, and with rooflights and other penetrations designed so they do not interfere with the drainage of the membrane.

Grade C – High humidity or special environments

Swimming pools and buildings containing liquids stored in open containers or where water is used in manufacturing, cleaning or storage processes, e.g. ice rinks, cold and chill stores, etc.

3.3.4.5 The BRE report *Thermal insulation: Avoiding risks – Report No. BR 262* recommends that built-up roofs with open ribs should always have a breather membrane to minimize the risk from condensation. This BRE report does not, however, consider the effect of internal environments or provide advice on making the breather membrane effective in low pitch roofs.

3.3.5 Vapour control and acoustic absorption

In some sports halls, swimming pool buildings, lecture theatres, etc., it is necessary to use a perforated lining sheet in order to absorb some sound energy in the soft thermal insulation. In this situation, insulation will be the first layer of product immediately above the liner, then the vapour control layer with a further layer of insulation above that completing the 'acoustic sandwich'. Good acoustic performance is usually only required in buildings of relatively high occupancy, and these buildings are likely to have higher than average levels of humidity. It is therefore particularly important that the vapour control layer should have a high vapour resistance, and its laps be well sealed.

3.3.6 Vapour control and rooflights

Rooflights of necessity penetrate the vapour control and, if fitted, breather membrane layers of any roof system. The continuity of vapour control layer should be maintained by fully sealing the rooflight liner sheet to the surrounding metal liner and/or vapour control membrane. Fasteners through the liners should have washers to seal the holes.

Breather membranes are intended to drain any condensate that forms to the eaves or trap it until it re-evaporates in better weather conditions. The breather membrane cannot be continued through rooflights and will not therefore have a drainage path to the eaves. Rooflights starting near the ridge and running in continuos lengths are therefore preferable to chequerboard layouts. Continuous rooflight layout should make allowance for safe passage on the roof past any long lengths.

3.4 SIDE AND END LAPS

3.4.1 Profiled metal external sheeting

The minimum end lap should normally be 150 mm for steel and aluminium profiled roofing, with the proviso that an increase to 200 mm is justified at movable end laps in aluminium sheeting when proprietary expansion-joint components are used (see Figure 3.5). Table 2.2 in Chapter 2 gives recommendations for end and side-lap joints.

For pitches down to 1.5°, a secret fix or standing seam profile should be used incorporating, if necessary, the manufacturer's recommended end lap details.

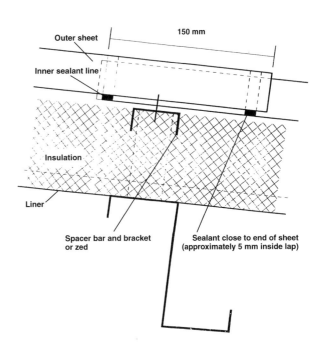

Figure 3.5 Built up insulated roof – end lap.

3.5 SECRET-FIX ROOFING SYSTEMS

3.5.1 Definition

3.5.1.1 The term '*secret-fix*' is used to embrace all systems capable of being erected without visible means of fixing and therefore includes 'standing-seam' construction. Self-supporting secret-fix roofing systems were introduced to reduce the risk of leakage at fasteners and side and end laps.

3.5.1.2 The use of panels extending from ridge to eaves in an unbroken length and a substantial depth of profile, provides the basis for leakage-free roofs with as little as 1° net pitch to give drainage.

3.5.1.3 In order to realise this net pitch, the structural design should be prepared with an increased fall of 1.5° above the minimum actual pitch.

3.5.1.4 The main problems to be expected from what is virtually a level surface is an accumulation of silt due to reduced water velocity and 'run-back' under the panels at the eaves under the influence of surface tension and aggravated by wind.

3.5.1.5 The latter issue is usually dealt with by turning down the panel edge and the inclusion of an eaves drip flashing or other similar measures.

3.5.1.6 Typical secret-fix systems have panels with side laps using one of the following methods:

- spring snap;
- mechanically seamed *in-situ* (often referred to as standing seam systems);
- batten cap over butt joint;
- panel engaged and rotated through 90°.

3.5.1.7 Not only is the system applied to panels of the very ductile aluminium but also to steel in a variety of pre-applied finishes.

3.5.1.8 The feature which distinguishes the standing seam from secret-fix design is the fact that the upturned longitudinal edges of adjacent panels are seamed together by a self-propelling seaming tool as virtually the final site operation. The seam may be formed to take in the panel fixing clips or it may be formed over fixing brackets or halters.

3.5.1.9 Either way, provision is made for the panels to move longitudinally over the fixing devices to ensure that thermal expansion does not overstrain what are quite rigid and substantial components. The thermal bridging effect is reduced by interposing non-metallic blocks between fixing devices and the structure.

3.5.2 Thermal movement

To avoid creep caused by thermal movements, a fixed point is introduced into the sheeting system. The method varies with the systems, but the fixed point should be designed to be strong enough to accommodate the thermal movement forces of the sheeting. Problems can arise where roof slopes intersect or there are penetrations through the roof as these should be designed to avoid the creation of any additional fixed points in the sheeting.

3.5.3 Wind forces

3.5.3.1 Wind forces on low pitch roofs are more severe than those on pitched roofs and are mostly in the form of uplift or suction forces. The self weight of most secret-fix systems is negligible and resistance to uplift is by the attachment strength at the clips. Wind pressures are most severe at verges, eaves, ridges and at features such as canopies. The attachment of the sheet to the clip or the resistance of the side lap to opening are most likely to limit spans under wind loads rather than the strength of the sheet itself.

3.5.3.2 For this reason the manufacturers' recommended spans should be followed and these should be based on testing of the complete system not individual components.

3.5.3.3 Secret-fix systems should not be used on curved roofs without reference to the manufacturers for their recommendations.

3.5.3.4 When considering secret-fix systems for ultra-low pitched roofs, specifiers should ensure that all necessary accessories are obtainable and compatible in design, quality and finish.

3.5.4 Site practice and installing

3.5.4.1 Secret-fix systems demand high standards of workmanship and supervision from initial setting out to completion, from operatives who have been given thorough relevant training. Typical tolerances are given below, however some systems may require smaller tolerances (see Figure 3.6).

3.5.4.2 Before a final decision is taken on the choice of a particular secret-fix system, the specifier would be well advised to consider the degree of difficulty involved, if it is necessary to remove a roof sheet or panel for repair or modification.

3.5.4.3 Most systems without secondary sealing strips are permeable to the passage of internal moisture vapour through the ribs, whereas systems incorporating sealant strips release internal moisture vapour to the atmosphere by passive ventilation longitudinally at the rib extremities.

3.5.4.4 The combination of low roof pitch and seam-jointing of the weathering sheets limits natural ventilation for the removal of moisture vapour. An efficient vapour barrier is therefore essential to minimize the risk of interstitial condensation.

3.5.4.5 A recent development is the use of a mobile roll-forming machine to produce very long secret-fix sheets on-site, saving the trouble and expense of delivering long loads. These may be used as self-curving sheets on shallow barrel vault roofs or curved on-site to tighter radii. The manufacturer's advice should be obtained at an early stage of design when considering site forming. Clear space with hard standing is at a premium on most building sites, but is essential if on-site manufacturing is to take place.

3.5.4.6 Further reading

MCRMA Technical Paper No. 3 – Secret Fix Roofing Design Guide.

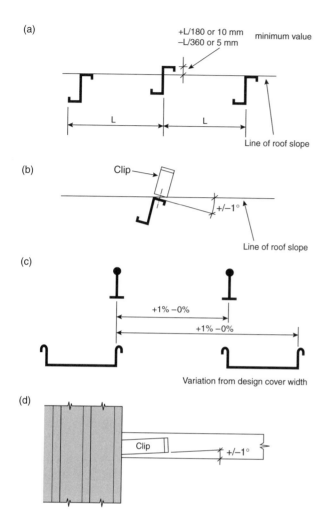

Element	Tolerances
Purlin levels	$+ L/180$ or 10 mm (*1)
	$- L/360$ or 5 mm (*1)
Purlins/clip slope	$\pm 1°$ (*3)
Clip spacing	$+ 1\%, - 0\%$ from sheet cover
	width (*2)
Clip alignment to ribs	$\pm 2°$

NOTES
*1 Choose smaller value.
  + = above design level, – = below design level.
  L = purlin spacing.
  Tolerance is in the level of a purlin relative to the purlins either side.
  There is a need for the eaves purlins to be accurately positioned relative to the rest of the slope.
*2 The spacing of clips can affect attachment strength, check manufacturer's guidance.
*3 The roof purlins should always be at 90° to the roof slope even if the slope is very low.
*4 These tolerances are for guidance only, check with manufacturer of the roofing system for any particular tolerance demands.

Figure 3.6 Installation structure tolerances (reproduced with permission from MCRMA).

3.6 CURVED SHEETING

3.6.1 Pre-formed curved sheeting

3.6.1.1 The scope for pre-finished, profiled sheeting in the construction of commercial and industrial premises, has been greatly enhanced by the introduction of new and technically challenging features. Such features include curved sheets produced from pre-coated material: curved external/internal mitred corners prefabricated and post-coated: and part spherical curves in GRP.

3.6.1.2 Whilst the roofing contractor is responsible for incorporating such details into the building neatly and efficiently, the designer is reminded that they demand greater accuracy from the structural frame and impose upon the sheeting contractor methods of working which lead directly to increased labour costs.

3.6.1.3 Curved sheets are normally produced from the same pre-finished profiled material as the remainder of the cladding, but its mechanical properties are changed due to the cold working involved in the various forming processes.

3.6.2 Reduced tolerance for fitting curves

3.6.2.1 One result is to stiffen the curved profile, limiting the degree to which its cover width can be adjusted to nest with adjoining sheets. It is essential therefore for the fabricator to maintain the same design tolerance on profile and pitch as in the basic sheeting.

3.6.2.2 The stiffer curve sheet can, especially with aluminium sheeting, result in bird-mouthing of the roof sheet lap. Additional care is needed at design, manufacture and installation to reduce this risk.

3.6.2.3 Double curves require even tighter tolerances than single curves. For this reason it may be easier to use single curves and lap joint on site.

3.6.2.4 If in addition to curved eaves and ridge sheets, the inflexibility of bonded or composite panels were to be introduced into the roof construction, the same high degree of accuracy would be required throughout.

3.6.2.5 The inclusion of a concealed gutter to create a break in the sheeting has advantages as it allows the wall and roof sheets to be fixed independently. Concealed gutters also minimize water flow down the curve and wall sheeting, which can lead to staining.

3.6.2.6 The length of straight tail on curved sheets must be considered at the design stage. Whilst for production purposes the fabricator will require a certain minimum tail length (possibly no less than 200 mm), a greater length would give the advantage of added flexibility for the nesting of end laps (see Figure 3.7).

3.6.2.7 It is generally agreed that a curved eaves sheet should extend up the roof slope only far enough to establish an end lap on the first purlin, but although a longer vertical tail would make it easier to plumb the sheet, it could also be argued that increased size and weight lead to handling problems. Thus for handleability it is recommended that the vertical tail be limited to give an end lap on the top rail.

3.6.2.8 The positioning of purlins and rails to support curved sections should be to the fabricator's recommendations.

3.6.3 Site work and appearance

3.6.3.1 When profiled sheeting is to be fixed vertically and continued across the roof unbroken by a gutter, inaccuracies in the geometry of the structural frame will be accentuated as the vertical profile is carried over into the roof line.

3.6.3.2 Furthermore, the design roof pitch must be accurately maintained if correct fitting of the curved eaves sheets is to be achieved.

3.6.3.3 Thus whenever features such as those mentioned above are to be incorporated, a thorough check should be carried out by the steelwork erector or main contractor prior to commencement of the sheeting operations, to ensure accuracy of the structural frame generally, correct positioning of rails and purlins, their alignment to a tolerance of 3 mm per 10 m run, absence of twist between planes, and roof slope $\pm 0.5\%$.

3.6.3.4 Accurate setting out is vital to the successful execution of schemes involving horizontal cladding or curved eaves sheets and a true datum is required on each sheeted face. Consideration should be given to the use of a template as a means of checking and adjusting the cover width of each sheet during fixing operation to prevent spreading.

3.6.3.5 Whenever special details such as GRP curves are required in non-standard colours or finishes, representative specimens cut from an approved sample supplied by the manufacturer should be held for reference purposes by the client, main contractor and sheeting contractor. These specimens would be used to determine the acceptability of materials in relation to finish, at the time of delivery.

3.6.3.6 Attention is drawn to the fact that profiled sheets from pre-finished material may have a surface 'grain' effect capable of creating an illusion of colour-change when viewed from a different angle. For this reason, care should be taken to avoid reversal of the sheets, even at a discontinuity where such action would otherwise go unnoticed.

3.6.3.7 Access to the working face is an important consideration in any cladding operation but particularly so when the curved sheets are to be fixed. The demands made on the fixers for accurate setting out, in handling the unsupported sheets and achieving correct alignment, call for access to the full length of the elevation during the entire fixing operation. It is extremely difficult and expensive to carry out these operations effectively from towers or similar equipment, piecemeal.

3.6.3.8 Fixing would normally commence at the down-wind end with the first wall sheet, followed by the first curved sheet located temporarily (but unfixed) and the first roof tier up to the ridge (secured temporarily). After adjustment of the eaves sheet and confirmation of line and squareness from base of cladding to ridge, the first tier would be fixed.

3.6.3.9 Subsequent progress would be made adopting a similar sequence one tier at a time, until perhaps four tiers are secured by temporary means. Not until these sheets have been adjusted should the permanent fasteners be inserted. Completion is thus achieved in increments of about 3 to 4 m, each stage involving fixing in two distinct operations.

3.6.3.10 When curved ridge sheets are included, work must proceed along both sides of the building simultaneously.

3.6.3.11 The use of corner sections requiring a reconciliation of profile suggests that sheeting should proceed away from the corners, meeting at an agreed point behind a site adjustable junction flashing.

3.6.3.12 Further reading
MCRMA Technical Paper No. 2 – Curved Sheeting Manual.

3.6.4 Ordering details

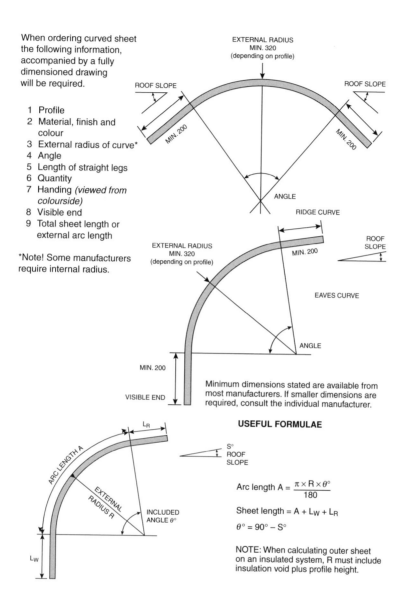

When ordering curved sheet the following information, accompanied by a fully dimensioned drawing will be required.

1 Profile
2 Material, finish and colour
3 External radius of curve*
4 Angle
5 Length of straight legs
6 Quantity
7 Handing (viewed from colourside)
8 Visible end
9 Total sheet length or external arc length

*Note! Some manufacturers require internal radius.

EXTERNAL RADIUS MIN. 320 (depending on profile)

ROOF SLOPE

MIN. 200

ANGLE

RIDGE CURVE

EXTERNAL RADIUS MIN. 320 (depending on profile)

MIN. 200

ROOF SLOPE

EAVES CURVE

ANGLE

MIN. 200

VISIBLE END

Minimum dimensions stated are available from most manufacturers. If smaller dimensions are required, consult the individual manufacturer.

USEFUL FORMULAE

L_R

ARC LENGTH A

EXTERNAL RADIUS R

INCLUDED ANGLE $\theta°$

L_W

S° ROOF SLOPE

Arc length $A = \dfrac{\pi \times R \times \theta°}{180}$

Sheet length $= A + L_W + L_R$

$\theta° = 90° - S°$

NOTE: When calculating outer sheet on an insulated system, R must include insulation void plus profile height.

Figure 3.7 Curves – ordering detail (reproduced with permission of the MCRMA).

3.7 HORIZONTALLY LAID CLADDING

The following check list has been developed in the NFRC Technical Bulletin 13, which covers design and workmanship of horizontal profiled metal cladding.

3.7.1 Sub-framing

1. Horizontal cladding emphasizes the defects in the line of the steel framework to a far greater extent than vertical cladding. The plane of the supports both horizontally and vertically should be straight to within a tolerance of 3 mm total in 10 m and there should be little or not twist between the vertical and horizontal planes.

Note: The tolerances required to achieve a 'fit for purpose' effect are small and special attention must be paid to the plan of the supports by the cladding contractor prior to accepting the sub-frame.

2. The recommended centres for supports should be as specified by the designer for the type of sheet specified. Recommendations for the width of supports will be determined by the method of jointing. In some cases, secondary steelwork may have to be provided.

3. Any fixings on the cladding face of the supports should be countersunk.

3.7.2 Sheeting

1. The choice of colour samples for profiled horizontal sheets should consider the various influencing factors which affect the perceived colour in use, e.g. building orientation, solar angle, product profile and texture. In general, horizontal cladding produces increased shadowing, i.e. darker perceived effect than that for vertical cladding.

   To minimize the risk of colour variation, orders for complete structures, or at least elevations, should be produced from the same production batch. This is particularly important for metallic and special colours.

   It is advisable for sections of typical quality and of the sample colour and texture, agreed with the client, to be cut into three pieces each to be retained by the client, main contractor and cladding contractor in the case of any dispute.

2. The manufacturer of the chosen materials should confirm the tolerances of sheet length, sheet width, gauge, and profile and the design of the cladding must incorporate allowances for these. The use of light gauge sheets must be considered against the increased potential sagging that may occur between supports, as it is unlikely that supports will be provided at less than 1.5 m centres. The same consideration must be taken against the use of shallow profile sheets which can also sag between supports.

   In some instances, roofing profiles have been specified which result in the side laps occurring on the profiled crest providing an unsightly detail.

   Many of the best results in all these instances have been achieved by the use of profiled sheet, minimum 0.7 mm thick, optimum range 32 to 38 mm deep and maximum length of 6 m with the profiled face outwards.

3. The numbers and positions of main and secondary fasteners will be determined by the structural and aesthetic requirements. It is generally recommended that a main fastener be placed at every sheet trough. Horizontal side laps should be sealed in normal circumstances. Where secondary fasteners to horizontal joints are necessary, the centres should not exceed 600 mm.

   Butt joints in horizontal sheets should be located adjacent to, but not over, the sub-frame support to avoid irregularities.

4. Vertical joints which are overlapped must allow for expansion of the underlapping sheet and consider the adverse visual effect of the vertical overlapping joints. Extra support steelwork must be supplied at these joints so that these expansion joints are accommodated.

5. Where vertical joints are required to be butted, the sheets can be laid with either staggered or straight joint. The latter can be made a feature by the introduction of a 'top hat' section (where it is necessary to establish the sheet length required) or by an 'open' joint of 6 mm between the sheet ends over a coloured profiled butt strap.

6. Where butt straps are used, these should be 200 mm wide and sealed to the sheet on both sides of the joint with two runs of appropriate mastic. They should be recommended at sheet end laps, as joints where four sheets meet are difficult to keep tight and therefore become unsightly. One drawback on the utilization of the butt strap system is the greater reliance placed on the accuracy of the design, manufacture and erection of the structural steel and sheet lengths.

7. Great care should be taken on setting out over large unbroken areas, as this may result in difficulties arising when closing or completing a 'wrap-around' cladding operation.

3.7.3 Access

Access to the full length of the face during the whole period of the fixing is necessary for setting out and handling of the sheets. Achieving correct setting and alignment from towers, etc. is extremely difficult and expensive.

3.7.4 Maintenance

1. As horizontal cladding is not self-cleaning by rain, it requires frequent washing to maintain its appearance. If this is not done, dirt becomes ingrained at the horizontal and vertical joint to the deterioration of the appearance of the building. This should be pointed out to the client in writing.

2. Horizontal cladding is particularly susceptible to damage by ladders that are used without spreader bars and the client should notify window cleaners, etc. where appropriate.

3. All cladding requires annual maintenance.

3.7.5 General

Good quality horizontal cladding is a combination of the correct quality and specification of the materials, correct design and detailing and good workmanship requiring the active co-operation of the designer, main contractor and specialist contractor. The observations for horizontally fixed cladding also apply where cladding is used to any angle other than vertically as on the chevron style.

4

COMPOSITE PANELS

4.1 INTRODUCTION

A composite panel is any laminated panel which is manufactured from different materials, permanently bonded together so that they act as a single structural element. For the purposes of this chapter, the description **composite panel** will refer to building cladding panels having metal facings to both surfaces, with an insulation core which completely fills the space between the two facings and is continuously and permanently bonded to both. The term used in European Standards for this type of factory insulated panel is 'Sandwich Panel'.

Site assembled systems using profiled insulation, or profiled sheets with insulation boards adhered to their troughs, are not composite panels by this definition. Such products should be treated as built-up metal roofing (see Chapter 3) except that they have the advantage of having no metal spacers which form thermal bridges. Where site assembled systems with profiled insulation are to be used the system manufacturer's recommendations for fixing and detailing should be followed.

4.2 MATERIALS

4.2.1 The metal facings

4.2.1.1 The metal facings may be boldly profiled, as for external roof sheets, or so lightly profiled as to be nominally flat. Almost flat facings are used for linings, and are often used for the external skin of wall panels. Fully flat facings are also possible, and are often used in modular panel systems.

4.2.1.2 Steel is the most popular facing material, as it provides the most cost effective panels for many applications. External facings usually have a thickness of

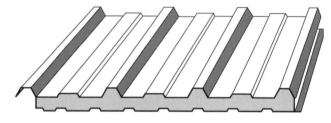

Figure 4.1 Typical profiled composite roof or wall panel (reproduced with permission of the MCRMA).

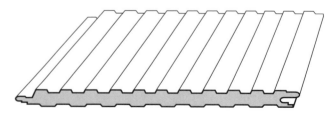

Figure 4.2 Typical composite wall panel (reproduced with permission of the MCRMA).

0.5 mm or 0.6 mm; linings are commonly 0.4 mm. The steel, similar to that used for built-up systems, is protected by zinc galvanizing, or aluzinc coating, with additional protection from a plastisol or other organic coating, see Chapter 2.

4.2.1.3 Aluminium facings are sometimes preferred in special applications; these could include very humid conditions, or a severe marine environment. The aluminium alloys are normally selected from EN 485 grades as follows: EN AW 3000 series, suitable grades are EN AW 3004 and EN AW 3105. External aluminium facings are typically 0.7 mm in thickness, and linings 0.4 mm or 0.5 mm.

4.2.1.4 Aluminium does not usually require coating for its protection, so mill finish preferably with embossed surface may be used. Where coloured finishes are required on aesthetic grounds, PVF_2 and abrasion resistant polyurethane are the popular choices for the external skin, while polyester is usually specified for linings.

4.2.2 The insulation core

4.2.2.1 The core material provides the thermal insulation, and contributes to the panel strength by providing most of the resistance to shear forces. The most common choices are polyurethane foam and isocyanurate foam both of which can now be produced 'CFC free' within the terms of the Montreal Protocol. Core thicknesses in the range 40 mm to 80 mm are normally required to provide U-values of 0.45 W/m^2 K to 0.25 W/m^2 K respectively.

4.2.2.2 Expanded polystyrene is sometimes preferred for flat, factory bonded panels. These usually need to be thicker than polyurethane cored panels, to achieve the same U-values.

Dense rigid mineral fibre insulation may be selected for applications where fire resistance, or acoustic insulation properties are considered to be most important.

4.2.2.3 It was traditionally thought that foam insulation lost some of its thermal characteristics through ageing; composite panel U-values are normally quoted with due regard to this loss of properties. Recent research has suggested that polyurethane foam, between metal facings, retains its insulation properties better than was previously believed.

4.3 MANUFACTURING

4.3.1 General

4.3.1.1 The manufacturing process involves the profiling of the precoated or mill finish metal skins, and their adhesion to the insulation core. The three basic methods are continuous production, batch production and individual panel production.

4.3.2 Continuous production

4.3.2.1 This is the normal method for high volume production. Manufacture commences with the facing materials being decoiled and profiled. The profiled facings are brought together and liquid state urethane is introduced between them. The urethane foams and expands until the panels reach their controlled thickness; at the same time, it bonds to both metal facings, by a process known as autohesive adhesion.

4.3.2.2 The foam cures as the monolithic panel advances down the production line. Tapes and vapour seals are adhered to the edges as the panel progresses.

4.3.2.3 The panels are cut to length as they reach the end of the line, usually by a form of travelling saw. Panel length is only limited by handling and transport constraints.

4.3.2.4 The panels are stacked ready for packing and delivery. The final stages of foam curing take place during the first hours in the stack.

4.3.2.5 Quality checks are possible at all stages of production, and the continuity of the process makes for consistent quality.

4.3.3 Batch production

4.3.3.1 Batch production was the standard production method before continuous lines were developed. The facings are profiled and cut to length (and this can be done in a different factory, or even a different country). They are then set up in static jigs or moulds. The liquid state urethane is introduced, and it fills the panels and bonds to the metal facings.

4.3.3.2 The panels must be left in the jig until the foam has cured sufficiently for them to be handled. They are then removed, and any necessary edge tapes and seals can be applied.

4.3.3.3 Manufacturing is much slower by the batch process than by the continuous process. Panel length is limited by the size of the jig. The method is now only used for low volume production, and for certain specialist panels.

4.3.4 Individual panel production

4.3.4.1 Individual panel production usually involves the use of an adhesive to bond the facings to the core. It is labour intensive and therefore relatively expensive.

4.3.4.2 The preformed skins are bonded to pre-cut insulation core material in some form of press.

4.3.4.3 This method of manufacture is often used to produce very flat modular panels, with pre-cut insulation cores and tension levelled metal skins. It offers extra flexibility in terms of panel width, profiling (if any), and product specification.

4.3.4.4 The size of the panels is limited by the size of the press.

4.4 PANEL PERFORMANCE

4.4.1 Structural

4.4.1.1 Composite panels are monolithic units in which the two facings act together with the insulation in resisting snow and wind loads, and foot traffic. Their combined strength and stiffness is far greater than the sum of the component parts – this is analogous to a lattice girder in which light angle sections are combined to create a stiff section which can bridge a large span.

4.4.1.2 The strength and stiffness of panels is determined by the facing material, its gauge, the core material and its thickness. Profiling of one or both faces can further increase the strength. A typical roof panel has a 0.5 mm steel outer facing and 0.4 mm steel inner facing, together with a minimum core thickness of 40 mm. Such a panel will typically span 2 m between purlins, under normal UK loadings; increasing the thickness of the core will permit the use of larger spans for the same loading conditions.

When aluminium facing materials are used, the thicknesses are increased to about 0.7 mm outer and 0.5 mm inner; these panels have similar spanning characteristics to the steel faced panels described above.

All panel manufacturers publish load/span data for their range of composite panel products. The safe maximum span of some composite panels, especially concealed fix systems, is often limited by the strength of the fastener and not the panel. This reduction in strength is caused by the point loading of the fixings

under wind suction loads. Designers should check that the design data used apply to the panel as installed and not a theoretical calculation or static load test on a well supported panel. If rooflights are included in the roof the maximum spacing between purlins will be limited by the strength of the rooflight panels.

4.4.1.3 Panels, with facings as described in section 4.4.1.2 above when fully fixed, can support normal foot traffic without damage. The foam core provides continuous support to the external facing to resist indentation. Personnel must wear suitable footwear to avoid marking or scratching the coatings.

4.4.1.4 Thermal bow effects can be caused when the facings are at significantly different temperatures, e.g. a roof panel in direct sunshine. The effect is accentuated when the external surface is a dark colour, and can be worse still for aluminium facings. Part of the design process should involve taking advice from manufacturers on limiting lengths, spans, colours, any special precautions, etc.

4.4.2 Thermal

4.4.2.1 Polyurethane and isocyanurate foam are extremely effective thermal insulation and, as they completely fill a composite panel, contribute to minimizing the overall thickness of panels. In a profiled roof panel, the insulation thickness varies, and the U-value is based on the average thickness. To achieve 0.45 W/m^2 K usually requires a 40 mm profiled roof panel (i.e. 40 mm thick at the troughs) or 45 mm flat wall panel. Composite panels insulated with other types of rigid insulation material, may need to be 60–70% thicker to achieve the same U-value.

4.4.2.2 There are very few cold bridges in composite panels; these are usually no more than through fasteners or fixing clips. It has been shown that these are unlikely to increase the U-value by more than 1–2%.

4.4.3 Condensation

4.4.3.1 Metal facings are effectively impervious to penetration by vapour, while polyurethane insulation has a closed cell structure which does not permit significant transmission of vapour. Interstitial condensation cannot occur without the presence of vapour in the insulation.

4.4.3.2 The joints, particularly at side-laps, require sealing to reduce the risk of condensation on the overlapping edge of the external facing. Most composite panels are supplied with a compressible seal in the side-lap, and this is adequate for 'normal' applications. Where the temperature or humidity is expected to be abnormally high, extra precautions may be necessary; the panel manufacturer will advise on suitable details.

4.4.3.3 When composite panels are used as cold or chill store insulation the vapour seals have to prevent inward moisture vapour pressure. Special treatments are necessary and the panel manufacturer's recommendations should be adopted.

4.4.4 Acoustic

4.4.4.1 Acoustic insulation properties are closely allied to cladding mass. Composite panels are relatively light and therefore do not have inherently good acoustic insulation properties. However, they can be installed with sealed joints to reduce airborne sound and very few paths through which sound travels easily. Such an installation is likely to perform almost as well as some built-up systems in reducing transmitted sound.

4.4.4.2 Acoustic absorption within a building depends mainly upon the nature of the lining. Composite panels have a nominally flat metal lining, which absorbs very little acoustic energy. Where sound absorption is required, it may be necessary to install additional acoustic lining systems.

4.4.4.3 Drumming from rain or hail on composite roof panels should be considered if occupants will be close to the underside of the roof.

4.4.5 Fire

4.4.5.1 Fire precautions in buildings are generally directed at reducing the risk of death or injury to occupants and the public and of economic loss. This is achieved by the selection of materials which provide required characteristics, and building elements which behave in a predictable manner. An explanation of fire performance ratings is given in Chapter 1.

4.4.5.2 Steel and aluminium liners achieve the highest classifications covered within BS 476 for combustibility, ignitability, surface spread of flame, etc.

4.4.5.3 Polyurethane and isocyanurate foam cores are charred by the application of heat to the metal skins. Polystyrene cores are not easily ignited behind the metal skins, but the polystyrene can melt and flow out of the panel and can then be a fire risk. Polystyrene cored panels should not be used for internal partitions and ceilings, or for applications with a high fire risk.

4.4.5.4 Fire resistant wall construction to BS 476: Part 20 is possible with steel skinned composite panels (aluminium has too low a melting point to be used for this purpose). Panels with isocyanurate foam cores have achieved over 4 hours integrity and 15 minutes insulation under these fire test conditions. They require special treatment at the joints, including cover strips, stitching fasteners, intumescent seals, etc. The manufacturer's advice should be sought for such applications.

4.4.5.5 Longer periods of fire resistance can be provided by composite panels with bonded, profiled, mineral fibre cores. Details of product specifications and test performance should be obtained from manufacturers. Users should remember that these panels will usually be thicker than polyurethane panels of similar U-value; this may affect sheeting lines, details, etc.

4.4.5.6 LPC and FM
Independent assessments of fire performances are available on some composite panels from LPC and FM insurer's organizations, see Chapter 1, sections 1.7.4 and 1.7.5.

4.5 STRUCTURAL

4.5.1 General

4.5.1.1 'Structural' matters include both the ability of the panel to resist the design loadings, and the structure of the building to which the panels are attached.

4.5.2 Design loading

4.5.2.1 Composite panels have to meet the same design performance criteria as other roofing materials, see Chapter 1.

4.5.2.2 Composite panels are subjected to thermal stresses, due to the 'thermal bow' effect described in section 4.4.1.4

4.5.2.3 Composite panel manufacturers publish tables or graphs for their products, showing maximum safe loads for a range of spans (purlin/rail spacing) and configurations (single span, multi-span, etc.).

4.5.2.4 Composite roof panels offer excellent resistance to damage by foot traffic, because the foam core provides continuous support to the outer skin, and helps to spread any applied point loads.

4.5.2.5 Composite panel manufacturers make allowance for the thermal bow effect when producing their technical literature. They publish advice on the effect of dark colours, and the possible restrictions in their use (e.g. reduced panel length, special laps, avoidance of some coating types, avoidance of south facing slopes and elevations, etc.).

4.5.3 Supporting structure

4.5.3.1 The panels are normally supported on purlins or sheeting rails; these must be of adequate strength and stiffness. It is the structural engineer's responsibility to design these supports, but the design process should take account of the stiffness of composite panels; the supports should not be unduly flexible relative to the panels.

4.5.3.2 Purlins/rails must be erected to a good standard of accuracy. Because the panels have high inherent stiffness, they cannot easily be deformed to follow uneven structures. The normal guidance for tolerance in the setting out of purlins/rails is that they should not be out of plane by more than one six hundredth of their spacing (e.g. ±3 mm for 1.8 m purlin spacing). This tolerance limit applies to the actual fixing of the panels. For the best aesthetic tolerances in high gloss or metallic finish wall panels, closer structural tolerances may be needed; the guidance of the panel manufacturer should be sought.

4.5.3.3 Roof panels are often required to have end-lap joints. At such joints, it is the external sheets which overlap – the joint in the lining and insulation is a butt joint (see Figure 4.3). As both sides of the joint require support, and there must be room for the fasteners at one side of the joint, it is essential that the purlins are at least

60 mm wide (and 65 mm is often preferred). When narrower purlins are used, it is possible to fit a plate to extend the width at the end-lap positions. These comments also apply to rails in cases where end-laps are required in wall cladding.

4.5.3.4 The alignment of purlins/rails becomes critically important at end-laps. If the purlins are displaced, or if they wander in their length, the lap may not coincide with the support. Bearing plates are the simplest solution in all but the worst cases.

4.5.3.5 Purlins/rails must provide adequate anchorage to the fasteners and must be capable of being fastened securely under site conditions. The minimum recommendations for support steelwork is given in Chapter 1, section 1.3.3.

4.5.3.6 It is a common requirement that cladding should provide restraint to the purlin/rail flanges. Composite panels are expected to provide such restraint where they are direct fixed to cold rolled purlins and light, hot rolled angles, channels and joists. Restraint is not provided by clip fasteners used for example with secret-fix systems, which allow panels to slide (e.g. under thermal movement). However, as some panels have fixings as widely spaced as one metre, or more, it is possible that this arrangement may not effectively restrain the flange, in such cases the purlin/rail manufacturer should be consulted.

4.5.3.7 Composite panels should not be used as a substitute for anti-sag bar arrangements since one function of the anti-sag bars is to hold the purlins/rails in their correct location, while the panels are erected.

4.5.3.8 Composite panels have a 'two way' rigidity, which is not shared by single skin profiled sheets. They are able to accommodate relatively large openings without the need for additional structural supports. As a guide only, holes up to about 350 mm diameter or 300 mm square, do not need structural supports or trimmers. Advice on specific arrangements should be sought from panel manufacturers.

4.6 ROOF APPLICATIONS

4.6.1 General

4.6.1.1 'Roof application' matters include those laps, seals, fasteners, flashings, etc. which are used in constructing a secure roof from composite panels. The traditional composite roof panel, with trapezoidal ribs, is through fixed and has simple side-laps. Concealed fix

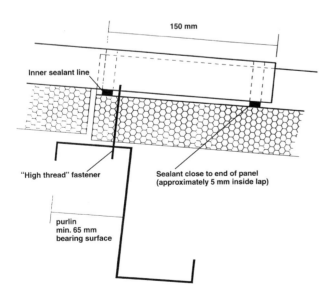

Figure 4.3 Through fixed composite roof panel – end lap.

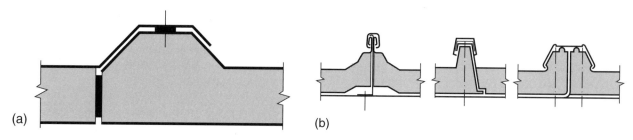

Figure 4.4 Side lap joints: (a) over 4°; (b) slopes greater than 1° (reproduced with permission of the MCRMA).

composite panels have more complex side laps and jointing systems, see Figure 4.4.

4.6.2 Traditional composite panels

4.6.2.1   End laps are introduced as required by slope lengths. The maximum length of panels is usually 10 to 13 m, a typical roof panel weight is 11 kg/m², so greater lengths may present handling problems.

4.6.2.2   The through fixings may be in the rib crown, the trough or, in some cases, a mini-rib within the trough. It is important to use purpose designed composite panel type fasteners approved by the panel manufacturer (as described in Chapter 7) in order to maintain the weather seal between the metal skin and the washer. Because metal skins are thin, the minimum washer diameter should be 19 mm in order to provide good pull-through strength.

4.6.2.3   The traditional arrangement is usually restricted to roofs with finished slopes of 4° or more. This is to ensure that run-off rates are reasonably fast, and that the through fasteners and laps will not be exposed to significant depths of water.

4.6.2.4   At end-laps, the lining and insulation is butt jointed over the purlin, and the overlap is formed in the external weather skin only. Manufacturers make their own recommendations, but it is usually appropriate to use a 150 mm overlap with two lines of sealant (see Figure 4.3). The sealant should be in preformed strips, and should be positioned at the top and bottom of the lap. Composite panels usually have very wide troughs, and the seals are most easily compressed by trough fasteners. In crown fixed systems, it may be necessary to use tail-bolting to compress the seals; this can be done by means of sealed rivets or stitching screws, there is no need to penetrate the panels fully.

4.6.2.5   At side-laps, it is usual to apply a single strip of sealant at the profile crown, this is likely to be 9 mm × 3 mm or similar. Side-laps are usually stitched at 450 mm to 600 mm centres according to pitch and exposure. Manufacturers publish recommendations in respect of side-lap stitching.

4.6.2.6   Side-lap stitchers can be either purpose designed stitcher fasteners or sealed rivets, see Chapter 7. The under lapping rib is very stiff, so it is simpler to make good side laps in composite panels than in single skin metal profiled sheets.

4.6.2.7   Care should be taken when fixing composite wall panels, especially aluminium or PVF₂ surfaced, to ensure that the fasteners are not over-tightened causing shallow dents around the fastener head.

4.6.2.8   Most composite panels have an integral side lap cover 'nib' within the lining sheet, and an integral, compressible strip within the insulation depth. This is intended to act as the vapour control and, provided the panels are fixed correctly, should be adequate for 'low risk' applications (i.e. normal temperature and relative humidity – Grade A environments, see Chapter 2, section 2.4.2).

4.6.2.9   In high risk applications such as food processing buildings, textile mills and indoor swimming pools, an additional seal is usually required at the lining. This is often achieved by means of a sealer strip at the liner side lap nib, and a continuous seal at purlins where end laps occur. Manufacturers issue guidance on any special requirements for their own particular products, which may include special coatings on the liner panels.

4.6.3 Concealed fix composite panels

4.6.3.1   The purpose of 'concealed fix' systems, is to conceal the fasteners from the weather so as to eliminate the risk of leaks at fasteners. This normally involves hiding the fasteners from view. Concealed fix systems are often used on very low pitches, sometimes as shallow as 1.5° finished pitch.

4.6.3.2 There are three common ways by which the fasteners are concealed. Crown fasteners can be covered by means of a separate 'crown cap' which is mounted on clips; simple versions are more effective at hiding the fasteners from view than rendering them weathertight. A refined version has fasteners at the side laps only and uses a close-fitting cap which is held in place by a spring action. This can give weather protection to the through fasteners. The third alternative is to eliminate through fasteners by the use of clips within the side laps; the joint may be further protected by means of a capping strip.

4.6.3.3 Logically, the elimination of exposed fasteners should be accompanied by the elimination of end laps. This can only be achieved by fitting the panels in single lengths from ridge to eaves, and this may involve the use of very long panels.

There is no doubt that such an arrangement can produce a roof which is both attractive and secure, but designers should be aware that the panels may become too heavy for manual handling (a 30 m panel is likely to weigh about a third of a tonne), and fixing may only be possible with a crane in continuous attendance.

4.6.3.4 Some manufacturers may allow the use of end laps in their concealed fix systems, subject to certain conditions or limitations.

4.6.3.5 In high humidity conditions, side laps and end laps (if any) should be given special attention as mentioned for the traditional arrangement.

4.7 STANDARD DETAILS

4.7.1 General

4.7.1.1 Flashings and fillers are generally similar to those used with built-up systems, but there are minor differences which may influence design choices, and special requirements that should be addressed.

4.7.2 Ridges

4.7.2.1 The spacing of ribs, in composite roof panels, is usually double that which applies in built-up systems. This means that the flashing must span far greater distances between supports (i.e. between profile ribs and especially at hip ridges). The flashing should be in an adequate thickness, which will be not less than 0.7 mm for steel or 0.9 mm for aluminium, and the edges should be stiffened with welts or feature steps.

4.7.2.2 The profiled fillers will be much wider than in built-up systems, and therefore more vulnerable to

being dislodged. They should be securely bedded in mastic and, in exposed high risk applications, other mechanical restraints should be considered; this could include retaining the filler with profiled metal section, especially on rake cut hips.

4.7.2.3 It is also possible to use crimp curved ridges, but this calls for very accurate working as neither panels nor ridges can be stretched to suit product tolerances.

4.7.2.4 The lining is usually closed with a metal trim mounted on the ridge purlins. Any gap between the ends of the composite panels must be insulated to eliminate cold spots or cold bridging. This can be done with mineral fibre, *in-situ* injected foam. In high humidity applications the liner trim should be sealed to the panels, and in its end laps, and injected foam preferred to mineral wool.

4.7.3 Eaves

4.7.3.1 Eaves panels are normally supplied with the foam and liner cut back, during manufacture, 50 to 100 mm from the end to promote water to drip off the panel into the gutter. A profiled foam or metal filler may be recommended to cover the exposed end of the insulation and metal liner, especially with mineral fibre insulation.

4.7.3.2 Exposed foam is degraded by solar radiation, and this leads to it becoming discoloured. The degradation is limited to a depth of a few millimetres and does not appear to affect the performance of the sheets. However, the discoloured foam can be considered unsightly; where the foam is on view, and as aesthetics are important, a cut back eaves detail is essential.

4.7.3.3 Crimp curved eaves sheets can be used to provide a transition from roof to wall. However, as mentioned for the ridge, these sheets are inflexible as are composite panels, so the detail calls for special accuracy on the part of the manufacturer and extra care and attention on the part of the fixers. Some designers prefer to introduce a break by including a gutter at the bottom of the roof slope.

4.7.4 Verges

4.7.4.1 Verge flashings should cover at least one rib of the roof sheets. The preferred arrangement for composites is to use Z or 'top hat' sections to support the verge flashing. The sections are bedded in mastic, and are usually located near the edge of the flashing.

4.7.4.2 The detail described above can also be used at parapets, or at steps in the roof line.

4.7.5 Openings

4.7.5.1 Small openings may be flashed by means of flexible EPDM soakers. These must be arranged such that no trough is completely blocked. An opening centred on a rib may be larger than one centred on a trough. Where the blocking of a trough is unavoidable, a special soaker upstand will be required and often structural trimmers are required to support the cut edges of the panels.

4.7.5.2 The simplest treatment for large openings with support trimmers is to fit a flat apron flashing over the rib crowns; this must extend from the ridge, past the lowest point in the opening. The down slope end of the apron flashing should be sealed across the trough using the method adopted at the ridge.

4.7.5.3 Specialist companies offer *in-situ* soaker kerbs in GRP, and these may be more acceptable, aesthetically, than large flat sheet arrangements. The GRP soakers are not structural and therefore support trimmers are essential.

4.7.5.4 Special factory fabrications can be made in aluminium. These can be welded and post coated to match the roof sheets. It is possible to use aluminium soakers with steel skinned panels, provided that the dissimilar metals are isolated by means of suitable paints, seals or tapes.

4.8 ROOFLIGHTS

Rooflight materials for composite panels are the same as those used for built-up systems, see Chapter 7. Double skin rooflights for use with composite panels, are factory assembled, and include some form of stiffeners or bulkheads to support the nominally flat lining sheet.

Composite panels, by virtue of their great strength and stiffness, are often used with relatively large purlin spacings. Rooflights which are not able to span such large spacings, usually control the purlin spacing, the normal maximum being 1800 mm. For this reason, independent barrel vault ridge lights are sometimes preferred.

4.9 WALL CLADDING

4.9.1 General

4.9.1.1 Any composite panels which may be used for roofing, may also be used for wall cladding. However, panels which are produced specifically for use on walls are designed with different priorities. It is generally simpler to make vertical planes weathertight, but aesthetic considerations are often paramount.

4.9.2 Roof systems as wall cladding

4.9.2.1 Roof panels usually have bold ribs, and these are sometimes valued for their contribution to the 'texture' of the wall surfaces. They can be fitted with the ribs vertical or horizontal. When panels are fitted with ribs horizontal, some thought should be given to the jointing of panels; normal side laps and end laps are functional, but end laps which are acceptable on a roof may be unattractive at close quarters. It is often better to use vertical 'top hat' flashings to make a clean break between successive panels.

4.9.2.2 Walls normally suffer far more penetrations than roofs. These include doors, windows, louvred ventilators, etc. All such openings must be trimmed and flashed, and these details can have a clumsy appearance due to the openings not being symmetrical in walls with bold ribs. Such problems may suggest to the designer that flatter composite panels be used.

4.9.2.3 Designers often want hidden fixings for wall panels. This may be essential for security, and is often preferable for aesthetic reasons.

4.9.3 Finishes for wall panels

4.9.3.1 Wall panels may utilize the same coating systems as roof panels.

4.9.3.2 Smooth and durable coatings such as PVF_2, which are not always suitable for roofs, are often preferred for walls. PVF_2 coatings have the added advantage of being available in metallic effects.

4.9.3.3 PVF_2 coatings are relatively thin, and can easily be damaged by careless handling. The appearance of a large area of high gloss panels can be badly marred by scratches and scuffs (e.g. from moving ladders against finished areas of work). For this reason, manufacturers usually offer a polyfilm protection for wall panel coatings; the film is stripped off once the panels have been fixed and are no longer at risk. The cost of this film is very small when set against the cost of replacing panels, which are rejected on grounds of surface marking or soiling. In general, these strippable films should not be left on fixed panels as they degrade quickly when exposed to weather and may become difficult to remove on completion of the contract.

4.9.4 Purpose made wall panels

4.9.4.1 Many designers, and building owners, prefer flat wall panels to the heavily ribbed panels which are used on roofs. Composite panels are ideally suited to the manufacture of such flat systems; they develop their strength from the sandwich of skins and insulation, and do not need profiling to contribute to strength or stiffness.

4.9.4.2 Continuously produced panels are usually made with a tongue and groove side lap detail which incorporates concealed fasteners. This is an aesthetically pleasing arrangement, but limits the number of fasteners which can be used. The rail spacing is often governed by the fastener strength. These panels cannot usually be end lapped, but are available in lengths up to about 10 m.

4.9.4.3 Continuously produced panels inevitably suffer very minor undulations in the metal skins; these arise from built-in tensions in the metal coil, tensions introduced during panel manufacture, and slightly uneven rising of the foam during curing. These tiny deviations would be invisible in a heavily profiled panel with a semi-matt, embossed finish; they could become all too prominent in a flat panel with a high gloss finish. Manufacturers address this possibility in two ways. They may increase the thickness of the outer skin (e.g. to 0.6 mm for steel or 0.8 mm for aluminium) and/or they introduce a very minor 'profile' into the outer skin, to conceal any unevenness. These 'profiles' are typically only about one or two millimetres deep. They can mimic a planked effect, or simply introduce mini-stiffeners at intervals of 50 mm or so. One version has a micro-sinusoidal profile, about a millimetre deep; from any normal viewing position, these disappear to give a completely flat appearance.

4.9.4.4 Panels with completely flat skins can be made (subject to manufacturing tolerances) by individual panel production. The insulation core is cut to thickness, so its surfaces are not subject to any variations in curing. The metal skins are cut from tension levelled coil and are often thicker than for continuously produced panels (e.g. 0.7 mm steel or 0.9 mm aluminium).

4.9.4.5 The size of individually produced panels is limited by the manufacturing facilities, they are usually supplied in lengths of up to about 4 m, and seldom more than 6 m. They are usually mounted in extruded aluminium mullions, and can sometimes have horizontal aluminium extrusions as well. The use of these extruded sections simplifies the incorporation of doors and windows and makes for very high aesthetic standards. These panels often depend upon highly efficient seals for their weather protection; these must be as durable as the panels themselves, EPDM or similar is preferred.

4.9.4.6 Manufacturers provide technical literature giving guidance on the design and fixing of their proprietary wall panel systems. Users should always take account of these published data.

4.9.5 Supporting structure

4.9.5.1 Section 4.5.3.2 gave guidance on the minimum requirements for tolerances in the setting out of purlins and rails. These suggestions were largely based on the structural requirements of 'stiff' panels; the section also mentioned that closer tolerances are required for the highest aesthetic standards in wall panels. This is a consequence of using nominally flat surfaces with high gloss finishes; reflections can exaggerate any deviations, in the same way that a reflection in a mirror can be grossly distorted by a minor deviation.

4.9.5.2 If high aesthetic standards are demanded from continuously produced, flat wall panels, the tolerances quoted in section 4.5.3.2 should be halved. Adjustable secondary steel supports are recommended, to compensate for normal structural steelwork tolerances. In the case of individually produced panels, fixed in mullions, it is usually possible to adjust the alignment of the extrusions to compensate for inaccuracies in the alignment of the steelwork.

4.9.6 Special details

4.9.6.1 Where roof panels are used as wall cladding, corners, jambs, cills, etc, are generally treated in the same way as for any other profiled sheets. If the panels are fixed with their ribs horizontal, it is often possible to use mitred panels at the corners; most manufacturers supply these as part of their package. Where these mitred corners are used, they should be of as small a girth possible; neither the panels nor the corners can be stretched to accommodate poor tolerances, so accurate setting out is very important.

4.9.6.2 Corner, jamb and cill flashings can also be made to suit nominally flat panels. These can usually be made to look neat and compact because there is no requirement to cover bold ribs or match profile modules.

4.9.6.3 Manufacturers of nominally flat, continuous production wall panels, usually offer a range of special panels to fit at corners and jambs. These include mitred panels and panels with curves to a choice of radii which can be made to suit horizontal and vertical applications. Users should remember that if a curved or mitred corner is to be used at both ends of an elevation, it is essential to incorporate a break within that elevation, to accommodate any tolerances.

4.9.6.4 Individually produced panels are normally supplied as a complete package which contains not only mitred and curved corners, but matching windows, doors, louvred ventilators, etc.

4.10 SITEWORK AND HANDLING

4.10.1 Site storage

4.10.1.1 Composite panels merit special care during storage, prior to fixing. Since panels are produced to order, any accidental damage, to either skin, can lead to whole panels needing to be replaced, which is expensive and may involve long delivery times.

4.10.2 Site handling

4.10.2.1 Packs may be lifted by forklift truck or crane (with a spreader beam where necessary). Where possible the packs should be hoisted on to the roof where they are less vulnerable to damage from other site activities. The structural engineer should be consulted as to safe positions and maximum pack weight.

4.10.2.2 Individual panels may be moved by hand – but panels must never be lifted by the overhanging end which has been prepared for an end-lap, as this can cause delamination of the outer sheet.

4.10.2.3 Panels should be lifted cleanly off the stack. Sliding a panel can damage the coating of either the facing sheet or lining.

4.10.3 Health and safety

4.10.3.1 All the normal safety rules and recommendations for building and roofing applications, should be followed, see Chapter 10. In particular, operatives should wear gloves to protect their hands against cuts when handling metal sheets. Also, goggles and dust masks should be worn during cutting operations with reciprocating saws (the use of abrasive wheels should be prohibited).

4.10.3.2 Operatives should receive instruction in handling heavy and awkward loads, a 12 m steel skinned panel is likely to weigh around 125 kg (aluminium skinned panels will be little more than half this weight).

4.10.3.3 Composite panel roofs contribute to site safety because, once fixed, they provide a safe working platform. They are fully walkable at all practical spans. Foot traffic on unfixed panels should be prohibited.

4.10.3.4 Rooflights are much lighter than composite panels and should be treated in the same way as rooflights in built-up construction.

4.10.3.5 Special care is required at end-laps to ensure that both panels have sufficient width of bearing on the purlin flange.

4.10.4 Fixing requirements

4.10.4.1 The normal requirements for fixing metal cladding apply to composite panels, although additional requirements are listed below.

4.10.4.2 Setting out of eaves and end laps must be very accurate, as discrepancies cannot be accommodated by varying laps.

4.10.4.3 Composite type fasteners approved by the panel manufacturer should be used, and these must not be over-tightened or they will dimple the thin skin of the panels. Correctly adjusted screw guns should be used.

4.10.4.4 Tail-bolting may be required to close end laps, especially in crown fixed panels.

4.10.4.5 Buildings where high humidity is anticipated should have their end laps bedded in mastic; side laps will also require additional sealant at the liner. These seals should be shown on the construction drawings, and must be continuous to be effective.

4.10.4.6 Where poly-film protection is used, it must be removed as early as possible to prevent it becoming strongly adhered through the action of UV radiation. In any case, the manufacturer's advice must be followed as to the maximum period before stripping.

4.10.4.7 Crown fasteners may pass through over 100 mm of panel thickness. Extra care is needed to ensure that they do not miss the purlins and are perpendicular to the surface.

4.10.5 Repairs to damaged panels

4.10.5.1 It is inevitable that a few panels will be damaged during construction or in service. Badly damaged panels should be replaced; this can be difficult due to seals and panel tolerances. Minor damage can often be repaired *in situ*.

4.10.5.2   A dent or split in the external skin cannot be left untreated as it will hold water, set up corrosion, etc. However, it is usually possible to repair damaged roof panels by over-skinning with a matching profiled skin. The edges must be sealed to prevent moisture getting between the two skins.

4.10.5.3   It is not usually possible to over-skin nominally flat wall panels, and these must probably be replaced. Where secret-fix details are used, with fasteners within the tongue and groove side lap, it may be necessary to remove panels from the end of the run (or top of the stack).

4.10.5.4   Scratches and abrasions should be treated before exposed metal starts to rust. Repairs should be made using the approved touch-up paint and a very fine brush.

4.10.6   Inspection and maintenance

4.10.6.1   The inspection and maintenance requirements of composite panels are generally similar to those for profiled metal cladding systems, see Chapter 10.

5

FIBRE-CEMENT SHEETING

5.1 INTRODUCTION TO FIBRE-CEMENT

Currently there are two patterns of fibre reinforced cement sheeting produced in the UK. A traditional 6″ profile has been widely used on industrial, agricultural, commercial, leisure and residential buildings for most of this century. It is defined by the pitch of its corrugations being 6″ and its thickness being a nominal quarter of an inch (in imperial measurements). The width of the sheet accommodates seven and a half corrugations. A corrugation at the end of the sheet is formed with sufficiently less depth than the others to permit the side lap with the next sheet to be formed evenly. The side lap for this profile is one corrugation. This 6″ profile is manufactured in, and widely available throughout, the UK.

The 3″ profile, the other fibre-cement corrugated sheet currently produced in the UK has eleven much shallower corrugations than the 6″ profile, but arranged at a three-inch pitch. The normal side lap for the 3″ profile is two corrugations. Historically this shallow profiled sheet has been used on smaller projects, on detached residential garages, portable buildings, the domestic out-building market and in the poultry rearing and laying industry. Light weight, ease of handling and simple fixing systems have created a market for this product.

Fibre-cement profiled sheets are manufactured in a continuous process from Portland cement, water and a reinforcing mixture of both natural and synthetic fibres.

This chapter describes only products made from asbestos-free formulations of fibres.

Manufacturers offer a conditional guarantee and claim that the durability of fibre-cement means that a life span of the sheet in excess of 30 years can be expected. Details of the specific guarantees on any project should be sought from the manufacturer.

Traditionally, all fibre cement profiled sheets have been classified as a fragile roofing material. Some manufacturers' 6″ profile sheets are now available with 'reinforcing strips' which can withstand the impact test described in HSE Special Inspectors Report 30. Designers should give serious consideration to their use since these reinforced 6″ profile sheets are intended to reduce risks during fixing and subsequent maintenance.

This chapter does not cover the application, installation or the health and safety requirements of profiled or flat asbestos cement sheeting. Instructions for the safe handling of asbestos cement sheets should be obtained from the manufacturer. The statutory health and safety precautions and requirements for anyone dealing with asbestos cement products to handle, cut, install, otherwise use, remove, or maintain it should be obtained from the Health and Safety Executive.

5.1.1 Appearance

A natural grey colour, determined by the colour of the cement used in manufacture, is available for all profiles.

Several cement sources means that minor colour shade variations are inevitable. A factory applied water based acrylic coloured finish is offered in a range of colours to suit most buildings and their locations. Advice should be sought from the fibre-cement manufacturers on the suitability, durability and light reflectance of colours being considered for projects by designers.

5.1.2 Properties

5.1.2.1 Standards for fibre-cement products

Profiled sheeting manufactured in accordance with the requirements of BS EN 494, shall have the minimum strength for profiled sheets given in Table 5.1. In addition, the sheets shall meet the permeability standard, frost resistance and durability standards before and after conditioning and exposure to soak/dry, freeze thaw and heat/rain cyclic tests. The manufacturer's literature should state a minimum density, e.g. 1450 kg/m^3.

Fibre-cement sheets are non-combustible to BS 476: Part 4 and in terms of external fire exposure they have a P60 (Ext SAA) rating to BS 476: Part 3 and may be classified Class O in accordance with the Building Regulations.

5.1.2.2 Physical properties

Water absorption

Average water absorption is up to 30% of dry weight after complete immersion for 24 hours.

Moisture content

When new, fibre-cement has a relatively high moisture content, if humid conditions prevail, damp patches (without formation of droplets) may appear on the under surface of the sheets. This phenomenon is not detrimental to performance and will disappear as a process of natural weathering.

Condensation

In the absence of proper ventilation, water droplets may condense on the underside of the sheeting, especially on cold clear nights when the roof radiates heat to the sky. Whilst not harmful to the sheet, adequate ventilation should be provided.

Effect of chemicals

Extremely polluted and aggressive atmospheres may cause a softening of the surface due to chemical reactions on the cement, and so advice should be taken from the manufacturer.

Biological

Whilst fibre-cement sheets are vermin- and rot-proof, moss or lichen may grow on the outer surface. This will not reduce the life of the sheeting. For advice on removal, please contact the manufacturer's technical department.

Thermal conductivity

Fibre-cement sheeting has only low thermal conductivity when compared with other sheet roofing products. This serves to reduce heat build up in summer and heat loss in winter. (Thermal conductivity, $k = 0.48$ W/m°C.)

Fittings

Fittings are handmoulded and should comply with the requirements laid down in BS EN 494. However, they should not be considered as load-bearing elements of a building's structure.

Light reflectance

Mean results for natural grey sheets are 40% dry and 16% wet, using magnesium carbonate as 100%.

Windloadings

When using profiled sheeting, the windloadings of a location should be determined so that the appropriate sealing and fixing requirements are selected.

Fibre-cement sheets should never be used as part of a 'stressed skin' construction.

The purlins to which the sheets are to be fixed should not deflect more than $L/200$, where L = the purlin span.

5.2 BRITISH STANDARDS FOR FIBRE-CEMENT PRODUCTS

5.2.1 Standards for profiled sheeting

Fibre-cement profiled sheeting should be manufactured in accordance with a quality system registered under BS EN ISO 9002 and to the Product Standard BS EN 494; Class 1X specification (see extracts in Table 5.1). The 3″ profile sheets typically fall under Category A whilst 6″ profile sheets typically fall under Category C.

The specifier should obtain confirmation that the fibre cement sheets being supplied comply with the required category e.g. Class A1X for 3″ profile or Class C1X for 6″ profile.

Table 5.1 BS EN: 494 – Strength categories

Long sheets (length > 900 mm)				
Category	Height of corrugation, s (mm)	Minimum thickness (mm)	Breaking load (N/m) Class 1	Bending moment at rupture (N m/m) Class X
A (e.g. 3″ profile)	15–30	4.0	1.400	40
C (e.g. 6″ profile)	40–80	5.2	4.250	55

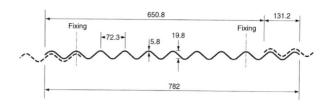

Figure 5.1 Fittings – 3″ profile (reproduced with permission of Eternit UK Ltd).

5.2.1.1 Sheet tolerances and labelling to BS EN 494

Length	± 10 mm
Width	+ 10 mm to –5 mm
Thickness	± 10% but not exceeding 0.6 mm
Squareness	Not to exceed 6 mm
Pitch	Category A ± 1.5 mm between crowns
	Category C ± 2 mm between crowns
Height	Category A ± 2 mm
	Category C ± 3 mm

5.2.1.2 BS EN: 494 – Requirements for labelling of fibre-cement products

A minimum of 50% of long profiled sheets and a minimum of 15% of short profiled sheets should be durably marked with at least items (a), (c), (d), and (e) from the list below. In addition, a minimum of 50% of fittings should be marked with at least items (a), (d), and (e).

a. The manufacturer's identification
b. The number of this standard
c. The category and class of the sheet
d. The date of manufacture
e. The material type (NT or AT)*
      *NT = New Technology, i.e. non asbestos
      *AT = Asbestos Technology

5.2.2 Fittings

The fittings and accessories should have a general appearance and finish compatible with the sheets with which they are to be used. Fittings are stocked in standard shapes and sizes (unlike metal flashings which are made to the customer's requirements). Figures 5.1 and 5.2 give typical designs; consult the manufacturer for availability of particular accessories.

Dimensional tolerances on fitting should be as follows:

Length and width:	± 10 mm
Thickness:	± 1 mm

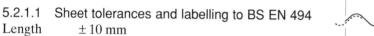

Figure 5.2 Fittings – 6″ profile (reproduced with permission of Eternit UK Ltd).

5.3. RECOMMENDATIONS FOR INSTALLATION

5.3.1 Movement joints

These are for use in long, continuous stretches of roof or vertical sheeting to provide for structural movement of the building. They are not normally required on buildings up to 45 m, unless special conditions apply. However, on longer buildings, they should be used as set out in Table 5.2.

Table 5.2 Recommendations for movement joints

Building length (m)	Number of movement joints
0–45	0
45–75	1
75–105	2
Every additional 30 m	Add 1 more joint

5.3.2 Lap treatments

Lap	This describes how much one sheet overlaps another at either the end (end lap) or the side (side lap).
Pitch	This describes the degree to which the roof slopes.

5.3.3 Design guidance procedure

Step 1: Exposure	Determining the expected degree of exposure by examining the map shown in Figure 5.3.

Step 2: Centres of support   Support centres for roof sheeting should be a maximum of 1375 mm for 6″ profile sheets or 925 mm for 3″ profile sheets, for a superimposed load of up to 1.89 kN/m^2. Where loads exceed 1.89 kN/m^2 contact the manufacturer for advice.

Step 3: Lap and seal   Establish requirement for lapping and sealing by reference to the map of the UK, Figure 5.3 and Tables 5.3 and 5.4. See later notes for more details.

Step 4: Fasteners   Calculate wind suction loading in accordance with BS 6399: Part 2 and check the number of fasteners required by reference to manufacture's advice.

Table 5.3 Sheltered and moderate exposure sites (less than 56.5 l/m^2 per spell)

Minimum roof pitch	End lap (mm)	Lap treatment	
		End laps	Side laps
22.5° and over	150	Unsealed	Unsealed
17.5° and over	300	Unsealed	Unsealed
15° and over	150	Sealed	Unsealed
10°–15° *	150	Sealed	Sealed
5°–9° †	300	Double sealed	Sealed

* The minimum roof pitch for 6″ profiled sheets is 5°, and for 3″ profile sheets, the minimum pitch is 10°.
† Where slopes are between 5 to 10° the maximum slope length should be 15 m with 300 mm end laps with a double seal and sealed side laps.

Figure 5.3 Categories of exposure to driving rain, derived from BS 8104; 1992 and BRE report *Thermal insulation: avoiding risks*, 2nd edn, 1994 (reproduced with permission of BSI).

Table 5.4 Moderate and severe exposure sites (more than 56.5 l/m² per spell)

Minimum roof pitch	End lap (mm)	Lap treatment	
		End laps	Side laps
22.5° and over	150	Unsealed	Unsealed
17.5° and over	300	Unsealed	Unsealed
15° and over	150	Sealed	Unsealed
10°–15° *	150	Sealed	Sealed
5°–9° †	300	Double sealed	Sealed

Note: On roofs over 10° pitch where parapets might allow snow build up, 300 mm double sealed end laps and single seal side laps are recommended.

* and †   See footnotes to Table 5.3.

5.4 INSTALLATION

5.4.1 General

The following guidelines should be observed to install profiled fibre-cement sheets.

1. The sheets should be installed smooth surface up.
2. The sheets should be cut with a hand saw or slow speed reciprocating power saw.
3. All fixing holes should be drilled, not punched, and provide adequate clearance for the fastener stem (minimum 2 mm).
4. There should be two fasteners per sheet width on every purlin or rail.
5. When using power tools in a confined area, dust extraction equipment is advisable.
6. The dust and swarf generated when working with the sheets does not require any special handling requirements other than normal housekeeping to maintain a clean working area.

5.4.2 Fixing

The correct fixing of a sheet is extremely important in order to avoid premature failure, corrosion or leaks in a completed roof. Many factors can influence the fixing of a roof, such as the type of purlin or rail and the nature of the roof in question. Of particular importance is the type of fastening system used and adherence to the manufacturer's recommended instructions.

The use of self drilling and self-tapping fasteners for crown fixing (e.g. those illustrated in Chapter 7), is recommended to fasten fibre-cement sheets. These special fasteners have been developed for fibre-cement sheeting and should be selected in accordance with the fibre-cement manufacturer's recommendations. A one step operation using these fasteners provides an automatic 2 mm clearance through the top sheet for the main fastener stem. It is important that the correct power tools, depth settings and recommendations are observed.

5.4.2.1 Method of fixing with proprietary self drilling fasteners

The fasteners are to be located on the crown of the corrugations in the standard positions recommended by the manufacturer.

The fasteners are to be installed at right angles to the purlins, never leaning up the roof slope. Therefore, the correct fixing position must be ensured by lining through with a chalk line or straight edge.

The appropriate fastener is only to be installed with the manufacturer's recommended power tool. Initial adjustment of the depth location is to be made as per the instruction manual.

The correct fastener is snapped into the nose piece of the power tool and the drilling point offered to the roof corrugation. The use of the machine's variable speed trigger will aid location of the fastener at the centre of the crown of the corrugation.

After initial location, the full speed of the machine is used to drill through the fibre-cement sheets as quickly as possible. When the drill point has passed through the sheets, less pressure is required to prevent damage to the drill point on contact with the steel. Less pressure is also required when drilling the steelwork. Excessive pressure at this time may damage the drill point.

The fastener is driven into the steelwork and the sealing washer compressed to provide full surface contact as illustrated in the power tool manual. The initial fastenings should be adjusted for tightness and depth location as required and once set up, should not need readjusting. Do not overtighten.

Visually examine installed fasteners to ensure continued correct tightness as illustrated in Figure 5.6.

After completion, if settlement of roof materials has occurred, re-tightening of the fasteners with hand tools may be carried out.

5.4.2.2 Method of fixing with hook or crook bolts and drive screws

Where the self-drilling type of fastener system is not being utilized, clearance holes must be drilled 2 mm larger than the diameter of the fixing. Drive screws should be central on the timber purlin.

Care must be taken to ensure that drills are kept perpendicular to the surface of the sheeting to prevent fixings being misaligned. All drilling swarf must be removed from the area of fixing.

The hook or crook bolts should be installed and appropriate washer and nut applied. Finger tightening of the nut should be used to ensure that the washer and

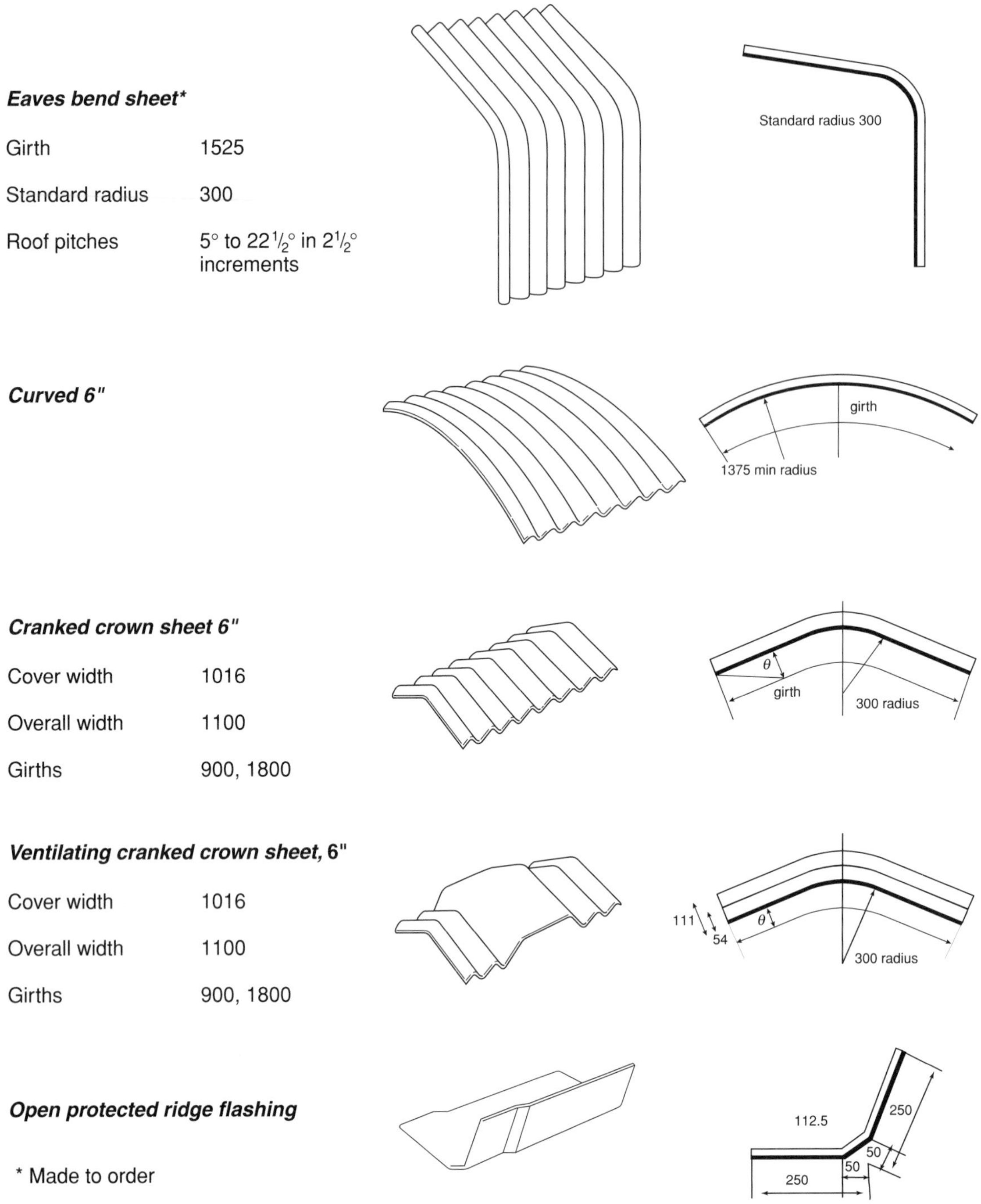

*Eaves bend sheet**

Girth	1525
Standard radius	300
Roof pitches	5° to 22$\frac{1}{2}$° in 2$\frac{1}{2}$° increments

Standard radius 300

Curved 6"

girth

1375 min radius

Cranked crown sheet 6"

Cover width	1016
Overall width	1100
Girths	900, 1800

θ

girth

300 radius

Ventilating cranked crown sheet, 6"

Cover width	1016
Overall width	1100
Girths	900, 1800

111

54

θ

300 radius

Open protected ridge flashing

* Made to order

112.5   250

50

50

250

Figure 5.4 Curved and cranked sheets, ridge fittings, flashings and closures; all dimensions are in mm unless otherwise stated (reproduced with permission of Eternit UK Ltd).

bolt is centralized on the clearance hole. Nuts should then be tightened until resistance is felt and washers/threads are not distorted.

Final tightening should not take place until an area of roofing is complete and the staging/scaffolding is to be removed.

5.4.3 Weather protection

All fixings should be protected from the elements with the appropriate washer and cap assembly. Proper protection at this stage will enable any future maintenance or removal to be performed easily and will, more importantly, protect the fixings.

Two piece close fitting ridge, 6"

Cover width	1016
Overall width	1124

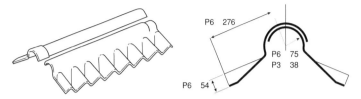

Two piece ventilating ridge, 6"

Cover width	1016
Overall width	1124

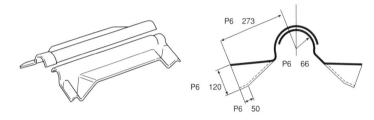

Two piece plain wing ridge, 6"

Cover width	1016
Overall width	1090

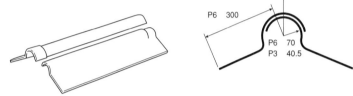

Two piece north light ridge, 6"

Cover width	1016
Overall width	1124

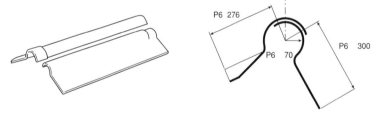

One piece plain wing ridge, 6"

Cover width	1080
Overall width	1200
300 wing angles	12½° to 45° in 5° increments
Roof pitches	15° to 45° in 5° increments

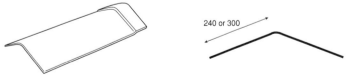

Figure 5.4 (Contd) Curved and cranked sheets, ridge fittings, flashings and closures; all dimensions are in mm unless otherwise stated.

Eaves filler piece, 6"

To close corrugations at eaves and provide a flat soffit for sealing at wall or gutter.

Cover width   1016

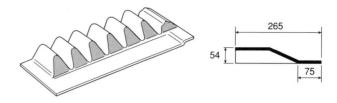

Eaves corrugation closure piece

Universal 6" fitting is illustrated.

Cover width   1016

Overall width   1055
                         65 to 150 backs

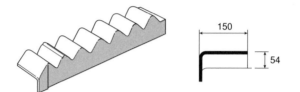

Apron flashing piece, 6"

Cover width   1016

Overall width   1086
                         LH only

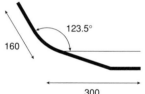

Under glazing flashing piece, 6"

Fitted over roof sheets below a run of glazing and for some details at head of vertical sheeting

Available left or right hand. Left hand fitting illustrated.

Cover width   1016

Overall width   1062

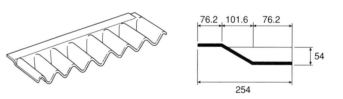

Figure 5.5 Closure fitments, all dimensions are in mm unless otherwise stated.

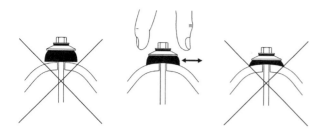

Figure 5.6 Method for checking tightness of fastener (reproduced with permission of Eternit UK Ltd).

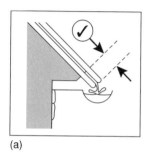 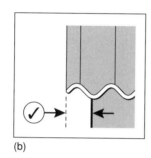

(a)   (b)

Figure 5.7 Correct positioning of (a) eaves; (b) verge (reproduced with permission of Eternit UK Ltd).

Where the building has a design life in excess of 25 years and/or corrosive processes or materials are present in the structure, consideration should be given to the use of stainless steel fasteners.

5.4.4 Overhangs

Sufficient overhangs must be allowed at the eaves to ensure that rainwater discharges into the gutter. The typical overhang into the gutter should be approximately 100 mm, or to 12 mm behind the centre line of smaller gutters.

Verges must be overhung by one complete corrugation unless a bargeboard is used.

5.4.5 Side laps

Where it is necessary to seal the end laps, and/or side laps, an extruded mastic sealing strip should be used. It is essential that the sealant is placed in the correct position to effect a weatherproof construction.

It is important to select a good quality sealant, see Chapter 8. Inferior sealants can lead to premature cracking, chalking and failure in use. For best results, choose a preformed, mastic ribbon of butyl or a poly-isobutylene-based material which has a rubbery, tacky characteristic and which will adhere to both surfaces when sheets are overlapped.

When appropriate, butyl strips should be positioned as shown in Figure 5.8.

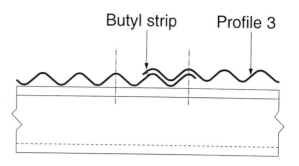

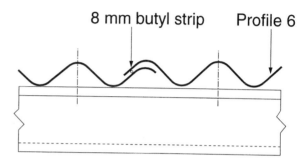

Figure 5.8 Side lap seal positions (reproduced with permission of Eternit UK Ltd).

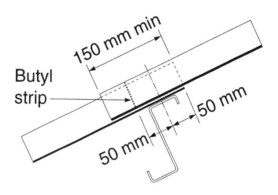

Figure 5.9 End lap seal positions (reproduced with permission of Eternit UK Ltd).

5.4.6 End laps

The minimum end lap for either 3″ profile or 6″ profile is 150 mm fixed as shown in Figure 5.9.

Where double sealing is necessary, the second butyl strip should be positioned 100–200 mm below the fastener. Recommendations for lap requirements are given under windloading (Tables 5.3 and 5.4).

5.5 FIXING PROCEDURE FOR LAYING WITH MITRED CORNERS

Unlike metal profiled sheets, fibre cement has a 5–6 mm thickness. Unless some corner mitring of the sheets takes place, the build up of sheet thickness at the lapping corners will affect the performance of the roof sheeting. The practice of 'box mitring' is not recommended as it will affect the waterproofing of the completed roof.

When laying and mitring fibre-cement sheets, appreciation of the different side corrugation heights on 6″ profile sheets is necessary, to achieve the correct 'nesting' and lay direction of the sheets.

1. Lay sheet No. 1 (eaves sheet) without mitring. Apply butyl strip as outlined by Tables 5.3 and 5.4.
2. Lay sheet No. 2, mitring bottom right-hand corner as per illustration, Figure 5.10.
3. Lay sheet No. 3, mitring as per Step 2; continue up roof to ridge.
4. Lay sheet No. 4, mitring top left-hand corner as per illustration, Figure 5.10.
5. Lay sheet No. 5, mitring **both** top left-hand and bottom right-hand corners as per illustration.
6. Repeat procedure from and including Step 3, working across the roof until there is room for only one more course to be laid, on the right-hand edge.
7. Lay sheet No. 7, mitring top left-hand corner and, if necessary, reducing the sheet width by cutting down the right-hand edge. (*Note:* To produce a weatherproof result, it is advisable to cut in the bottom of the valley of the sheet rather than at the top of the ridge, to leave a down turned overhang.)
8. Lay sheet No. 8 as per Step 7.
9. Lay sheet No. 9 without mitring.

Note: 1. On a duo pitch roof, start both slopes from the same end of the building.
             2. Corrugations of sheets must line up at the apex to ensure that the ridge accessories will fit.

5.6 NOTES FOR GUIDANCE ON STORAGE

5.6.1 Natural grey sheets

1. When handling sheets, lift by the ends only.
2. Sheeting stacks should always be raised above damp site conditions, they should generally not exceed 1200 mm high unless a flat concrete base is available, when the maximum height can be increased to 1500 mm.
3. If stacked on bearers use at least two bearers for sheets up to 1.5 m long and at least three bearers for longer sheets to raise them off the ground. Never stand or stack fibre cement sheets vertically on their edges.
4. Protect sheets from damage. Store as close as possible to fixing site, allowing room for handling.
5. Stack smooth face up.
6. Protect from wind by stacking in a sheltered position or holding down top sheets with ropes, weights or clips, see Figure 5.11.

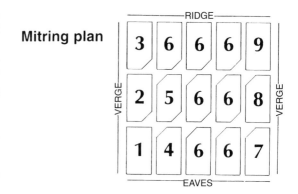

Mitring plan

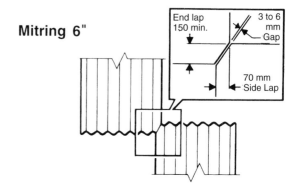

Mitring 6″

Figure 5.10 Mitring guide (reproduced with permission of Eternit UK Ltd).

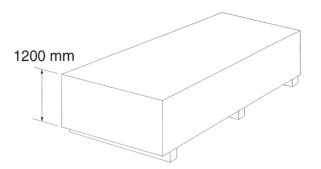

Figure 5.11 Storage of sheets.

7. A separate stack should be made of each length of sheet. If this is not possible, stack with the smallest on top and the longest on the bottom.
8. The top two sheets of each stack can also be left projecting for additional weather protection.

9. Note that after installation, owing to the vapour permeability of fibre-cement profile sheeting, dampness may appear on the underside of the sheet. This is a temporary phenomenon and will disappear following successive wet and dry periods. It in no way affects the weatherproof quality of the sheets.

5.6.2 Coloured fibre-cement and metal lining sheets

All previous storage recommendations for natural grey sheets apply.

Additionally, factory coloured sheets should be stored under cover at all times, preferably in a building. If inside storage is not available, then a tarpaulin or similar can be employed. If a tarpaulin is used, provision must be made for effective air circulation around each stack. This includes spacing the tarpaulin off the top and sides of the stack to avoid condensation.

Ingress of moisture into coloured profile sheets at this stage may detrimentally manifest itself later in:

- efflorescence staining;
- bowing during installation;
- permanent distortion;
- rust stains (of metal lining sheets).

If the sheets are to be site painted, ensure that the paint used is vapour permeable.

Warning: Never seal one side of a fibre-cement sheet without also sealing the reverse side, otherwise the moisture movement of the sheet will be affected and this will give rise to cracking.

Never directly apply insulating foams to the underside of fibre-cement profiled sheeting, as this changes the characteristics of the sheets and severely restricts any independent movement between the fibre-cement sheets and the structure. This will result in overstressing and cracking of the roof sheet.

5.7 SAFETY

5.7.1 Preparatory safety

- The structure should be adequately prepared for the sheets.
- Nobody should be allowed directly beneath the area being sheeted.
- Always treat fibre-cement as a fragile roof.
- Workers on a roof should wear suitable clothing, appropriate safety boots or safety shoes (not Wellington boots), and avoid loose, flapping clothing.

- The position and fixing of all purlins and rails should be checked before commencing sheeting.
- Ensure that there is proper access up to the roof.
- Provide a scraper at the bottom of all ladders to remove mud from boots.
- Crawling boards should be used for roof access.

5.7.2 Safety at work

Fibre-cement is classified as a fragile roof.

Workers must not be allowed to use the roof as a working platform during sheeting. Under no circumstances should the roof sheet be accessed or walked upon without the correct crawling boards, ladders, and other safety precautions. Fibre cement sheets should never be directly walked upon as this can cause stresses and cracking. Materials should not be stacked on the roof.

On calm days, it is possible for one person to safely handle smaller sheets at roof level. In other circumstances, safe handling of profiled sheets on a roof requires two people. Two roofers are always required to lay the eaves course and the sheets above rooflights.

Always lay the sheets in accordance with the approved sequence.

Do not cut the sheets in a confined space since nuisance dust will be created. This dust is classed as nuisance and non hazardous (refer to product data sheets – Health and Safety). Preferably, sheets shall be cut at ground level on suitable rigid supports using hand or powered saws. Powered saws should be of the reciprocating saw type and **not** disc or circular blade devices. Experience has shown that hand or powered saw blades having 3–3.5 mm teeth pitch are most suitable.

Remove all loose material from the roof as the work proceeds.

Always fully fix the sheets as work proceeds.

Do not leave tools on the roof's surface.

Avoid deflecting a sheet whilst attempting to force a bearing.

Sheets should be laid in tiers from the eaves to the ridge, thereby allowing easier use of crawling boards. Correct staging should always be laid over the purlins ahead of the sheeting. Heavy loads, such as industrial ventilators, must not be carried over the sheeting without the use of crawling boards.

Where regular access is required to reach rooflights, ventilation and service ducts, properly-constructed walkways should be provided.

Take extra care on a roof during windy, wet or frosty weather. Do not step on side lap corrugations. Take extra care on painted sheets whose surface will be more slippery than natural grey sheets.

Always observe the relevant provisions of the Health and Safety at Work legislation currently in force.

5.8 ROOFING SYSTEMS

As well as single skin roof constructions, it is quite common to use fibre-cement roof sheets in combination with lining sheet and insulation materials to provide a thermally efficient roof. There are three principal constructions that are suggested, although for special applications, such as swimming pools and laundries, the manufacturers can develop variations to suit high humidity conditions.

5.8.1 Metal lining system

A pre-finished metal sheet is laid on the roof purlins and fixed with a proprietary metal spacer system. In the depth provided, insulation quilt is rolled out and neatly laid around and under the support system. Finally the weathering fibre-cement sheet is laid as previously described and secured with top fix devices to the support system.

Typically, 83 mm thickness of insulation quilt will provide a U-value of 0.45 W/m K when calculated using the BRE method for a metal spacer bar, as described in Chapter 3.

5.8.2 Rigid insulation boards

Various types of pre-finished and natural rigid insulation boards can be laid on the purlins prior to fixing the weathering fibre-cement sheets. The insulation boards are installed with horizontal joints on the purlins and galvanized inverted 'T' sections used along their long edges. Manufacturers of these insulation boards can supply detailed information on the correct installation and the need to seal or tape the board joints.

As fibre-cement sheets are relatively heavy, they can impose significant line loads on the insulation boards particularly over the purlin top flange. In addition to the dead load of the sheet, continual wind pressure and suction together with occasional snow loading can greatly increase this line load on the insulant. Therefore it is essential that a suitable load distribution strip is placed above each purlin line on the insulant prior to laying and fixing the fibre-cement sheet.

Typically 40/50 mm thick insulation boards can provide a U-value of 0.45 W/m K.

5.8.3 Double skin system

To maximize the unique characteristics of the material, a fibre-cement liner sheet is made in the same sizes as the weathering sheets. This liner is laid on the purlins and a timber batten above this sheet is used to form an insulation space. Once the insulation quilt is rolled out and trimmed around the liner and batten, the weathering sheet can be installed as previously described.

Typically 80 mm thickness of insulation quilt will provide a U-value of 0.45 W/m K

With all these insulated systems the manufacturers can offer information and advice on the suitability of the roof pitch, condensation risk, practical installation, rooflight insertion, etc.

5.9 MATERIALS AND PERFORMANCE

5.9.1 Material characteristics

Installed weights

For design purposes the following as-laid weights should be employed:

Single skin fibre-cement	17 kg/m^2
Insulated metal liner system	22 kg/m^2
Rigid insulation boards	18 kg/m^2
Double skin fibre-cement	28 kg/m^2

5.9.2 Chemical resistance

Over years, chemical and industrial atmosphere pollution will cause a slight softening of the surface of natural finish fibre-cement profiled sheets. The acrylic paint finish provides added protection against many acids, alkalis and solvents normally found in the atmosphere.

The polyester/alkyd finish to the metal liner tray is corrosion resistant in normal internal industrial atmospheres. However, in more aggressive environments, a fibre-cement liner tray is recommended by the manufacturers and, in particularly aggressive environments, the manufacturer should be contacted for advice.

5.9.3 Thermal movement

Thermal movement is negligible, but it is necessary to provide movement joints in association with the structural framework.

The coefficient of linear expansion is 8×10^{-6}.

When roofing buildings where there will be higher than normal temperatures (e.g. foundry or kiln type buildings), cement based sheets should be stored near a heat source prior to fixing and be fixed on a dry day. The necessity for providing additional movement joints* over areas which are subjected to sudden changes in temperature must be considered.

*Consult the relevant manufacturer for recommendations.

5.9.4 Sound insulation

When tested in the frequency range 100 to 3150 Hz , the roofing systems should achieve a mean airborne sound reduction as follows.

Single skin fibre-cement	26 decibels
Insulated metal liner system	38 decibels
Rigid insulation boards	28 decibels
Double skin fibre-cement	41 decibels

5.9.5 Condensation

It is important to note that although the sheeting itself is watertight, it has the ability to absorb moisture and dissipate it in more favourable conditions. This material characteristic has a significant effect in reducing condensation occurrence.

With all joints taped and/or sealed the metal lining tray and polyisocyanurate insulation can act as vapour control layers, reducing airborne vapour and interstitial condensation risk.

5.9.6 Effect of frost

Fibre-cement sheeting is not affected by normal frost conditions experienced in the UK.

6

ROOFLIGHTS

6.1 INTRODUCTION TO ROOFLIGHTS

Rooflights are required in most industrial and commercial buildings to create an acceptable working environment. The HSE Workplace (Health, Safety and Welfare) Regulations 1992, Regulation 8 states the following.

1. '*Every work place shall have suitable and sufficient lighting.*
2. *The lighting shall, so far as reasonably practical, be by natural light.*'

With profiled cladding, the commonest form of rooflight comprises matching profiled translucent sheets – usually double skin. Alternatively barrel-vaults or domes can be fitted on up-stands, on both sloping and flat roofs.

Profiled rooflights have been used successfully for nearly 50 years and, provided they are properly selected, specified and installed can remain serviceable for 30 years or more. Manufacturers guarantees of 15–20 years are available. Careful selection of high quality materials and correct fixing is of critical importance to durability. Follow the recommendations of the rooflight manufacturer.

Rooflight solutions are available to meet every application, and to fulfill every requirement, including CDM, fire resistance, loading and spanning. Experience suggests that the best results are obtained if the rooflight requirements are developed at an early stage in the design process. BS 4154: Part 1: 1985 provides a standard for GRP rooflights, and BS 4203: 1980 provides a standard for PVC. No standard yet exists for polycarbonate. BS ENs are in course of preparation by the European Committee for Standardisation (CEN).

6.2 DESIGN ASPECTS

6.2.1 Material and types

6.2.1.1 Material
Rooflight materials fall into two groups:

1. thermosetting, GRP (glass-reinforced polyester), which do not soften with heat;
2. thermoplastics, including PVC, polycarbonate and others which soften with heat.

In general, GRP rooflights may be used at greater spans than PVC and polycarbonate.

Rooflights are available in a range of fire resistant qualities to satisfy the requirements of the Building Regulations. Thermoplastic rooflights have more limited applications, see section 6.2.5 for full details.

It is generally found that the thermoplastic rooflights need extra attention when being fixed.

All rooflights are available in a range of qualities, thicknesses and surface coatings. The thickness is varied according to structural requirements, safety requirements and imposed loadings.

GRP is available from 1 mm (1.83 kg/m^2) up to sheets over 3 mm thick, the latter having significant structural rigidity. PVC and polycarbonate are generally 1.3 or 1.5 mm thick. Thinner rooflights, e.g. 1.53 kg/m^2 GRP, are not recommended.

Surface coatings provide different levels of durability, chemical resistance and weather resistance. For GRP the choices range from polyester film, polyester gelcoats to more expensive PVF film.

6.2.1.2 Type

Rooflight installations usually fall into one of the following types:

1. single skin (see Figure 6.1);
2. double-skin, site assembled (separate liner and weather sheet) (see Figure 6.2);
3. double-skin, factory assembled (installed as a single unit) (see Figure 6.3);
4. continuous vaults, or individual domes, installed on upstands above the roof (either factory assembled, double-skin site assembled or single skin);
5. multi-skin rooflights, with three or more layers, may be specified for improved strength and insulation.

Selection depends on the application. Types 1, 2 and 3 are suitable for all metal and fibre-cement profiles, down to pitches (roof-slopes) of about 6°. Vaults and domes on upstands are more suitable for low slopes, below 6°, and flat roofs.

Factory assembled double-skin units are typically installed with composite panels, whereas site-assembled rooflights are preferred with built-up roofs.

6.2.2 Daylighting levels

Daylight is subjective, and quality, i.e. glare levels, diffusion and location, is more important than intensity. Given a reasonable distribution of rooflights, satisfactory results can be achieved with rooflight areas ranging from 5% of floor area for warehouses, up to 15% for industrial buildings, and 20% for sports halls and factories where intricate work is done, or for urban areas where air quality/pollution reduces the level of natural light. 10–15% proves satisfactory for most industrial premises. The proportion of rooflights may have to increase slightly where rooflights are specified in a heavily diffusing finish or when they are concentrated, e.g. in ridge to eaves applications. Typical light transmission for double-skin rooflights is around 70% when new. Diffusing agents can be added during manufacture, to minimize glare, solar gain and deep shadows.

All rooflights perform better and last longer if they are kept clean. Washing down with water and mild detergent, at 1–2 year intervals, will maintain maximum

light transmission. Dirt deposits allowed to build up over years will shorten the life of rooflights.

6.2.3 Weathering and durability

Plastic rooflights are generally very resistant to normal pollution in the atmosphere provided the products have been protected with ultra violet light (UV) inhibitors, and suitable surface protection. With the use of special coatings and films, the products can be suitable for very aggressive chemical environments and often will retain structural performance for longer periods than metal sheeting of a similar profile.

Surface protection can be either liquid applications (gel-coats and lacquers), or films (polyester and PVF) or co-extruded layers on thermoplastics. Resistance to discoloration, surface degradation and embrittlement depends to a large extent on the protective treatment used by the manufacturer.

UV attack depends on location, on orientation of the roof relative to the sun, and on the intensity of sunlight. Advice on UV resistance is available from manufacturers based on their accelerated weathering tests. GRP, with good UV inhibitors and a polyester or gel-coat protection, and kept reasonably clean, can be expected to remain structurally sound for up to 30 years. Discoloration can begin within about 5 years, depending on weather, but reasonable translucence should be maintained for over 10 years. Performance thereafter depends on environment, quality of sheet and maintenance (see Chapter 10). Higher fire-resistant sheeting exposed to UV light discolours more quickly, due to the fire-retardant additives.

Manufacturer applied gel-coats generally extend first maintenance to 12–20 years, dependent on the type of gel-coat. Gel-coats are recommended where very chemically aggressive environments exist due to industrial processes or marine locations. PVF film surfaces are available which are very weather-resistant and extend the daylighting life up to 30 years before maintenance.

Polycarbonate must have surface protection, otherwise it is attacked by UV light. With good protection, a mechanical life of 20 years may be expected, with transparency retained for at least 10 years.

Some PVC darkens and embrittles under UV light. A useful life of 5–10 years has been common for such grades. However, specially formulated and protected grades are available, and if properly fixed, can last up to 20 years.

6.2.4 Typical application

There is a wide choice of cladding systems on the market; the type of system influences the choice of rooflights.

For built-up roof systems, comprising separate liner and outer sheet, of metal or fibre-cement, secured with through fasteners, site assembled double-skin rooflights are recommended. This allows for easy ridge-to-eaves ventilation; provided liner panel joints are all sealed and taped, condensation is unlikely.

With built-up roof systems comprising weather sheet and insulation board, it is preferable to fit a translucent weather sheet above a flat double-skin factory assembled rooflight box. This gives a triple skin rooflight construction which maintains ridge-to-eaves ventilation and ensures it is easy to maintain a vapour check. Although a traditional method of insulating a building, the use of insulation boards in 'T' bars presents special problems with safety, as the fixed liner system is designated as fragile.

Factory assembled rooflights are suitable for composite cladding panel systems; such rooflights should fit closely to the adjacent composite panel to provide a vapour check. Ventilation of rooflights in this form of roof construction is neither necessary nor practical.

With 'secret-fix' systems, profiled rooflights with fixings through the sheets are sometimes used. With very low roof slopes great care is needed in the installation to ensure effective lapping and sealing; in these applications the use of barrel vaults or domes on upstands are preferable.

6.2.5 Building Regulations

6.2.5.1 Fire resistance

At the time of publication, the fire performance requirements for external roof and wall surfaces, and internal linings, are defined in Approved Document B (1992 edition) (AD B'92). At the time of writing, the Department of the Environment, Transport and the Regions (DETR) has out for consultation some proposed amendments to AD B.

Fire performance requirements vary according to building type, size and proximity to boundary, location of lights within the roof or wall, and with proximity to common areas, escape routes and internal fire walls. In addition, there are variations in Building Regulations between England, Scotland and Northern Ireland.

However, for a typical industrial building, less than 20 m high and more than 1 m from any boundary, and excluding escape routes or areas within 1.5 m of a fire wall, the requirements are usually as follows.

Outer surfaces (roof and wall)

The requirements for the outer surface are generally fire penetration requirements (e.g. AA or AB) to BS 476:

Part 3: 1958. For pitched roof applications a sloping test should be carried out, and the rating prefixed S (i.e. SAA or SAB).

There is no restriction on the use of a material which is rated AA, AB or AC on the outer surface of a roof.

The requirement for walls is generally similar except that walls within 1 m of a boundary should be rated Class O (as defined in the Building Regulations), and walls more than 20 m tall should have a fire propagation index less than 20 (as defined in BS 476: Part 6: 1989) for the first 20 m from the ground, and be rated Class O for the higher areas.

Linings

The requirements for linings are generally surface spread of flame requirements (e.g. Class 1 or Class 3) to BS 476: Part 7: 1987. There is no restriction on the use of materials rated Class 1 as the lining of the walls and ceilings, except for 'circulation spaces' and within 1.5 m of any compartment wall.

A reduced rating of Class 3 is acceptable, but only if each rooflight or group of rooflights has a maximum area of 5 m², with a minimum dimension of 3 m in any direction between each rooflight group.

A typical rooflight should therefore have an outer skin rated SAB and a liner rated Class 1.

Note: Single skin rooflights form part of the outer skin and the liner, therefore they should meet the requirements for both (with the reduced requirement for liners applying where appropriate).

These requirements apply to most normal industrial buildings, but there can be variations and it is essential that the designer/roofing contractor specifies the correct fire rating for any given application, as defined by the applicable Building Regulations.

GRP rooflights are available in all of the above grades. Thermoplastic materials (e.g. PVC and Polycarbonate) cannot be tested to BS 476: Part 3, but can be deemed to be rated AA if they are rated Class 1 to BS 476: Part 7. Other materials may achieve lower ratings of Tp(a) or Tp(b); there are additional restrictions on the use of these materials: for example, they should not be used within 6 m of a boundary.

6.2.5.2 Thermal insulation

Approved Document L 1995 defines that up to 20% of the roof area of industrial buildings can comprise rooflights with a U-value of 3.3 W/m² K (i.e. double-skin); it also defines three possible methods of deviating from this basic requirement. Multi-skin rooflights with more than two skins can achieve lower U-values.

Table 6.1 Maximum permitted area of roof as rooflights

U-value W/m² K	2.0	2.5	3.0	3.3	3.5	4.0	4.5	5.0
Max. permitted area of rooflight %	37	28	22	20	19	16	14	13

In the elemental method, the rooflights can be considered independently, and the area can be varied to reflect varying U-values as summarized in Table 6.1.

The approved document also includes a calculation method (which allows variation in rooflight area to reflect variations in the insulation of other parts of the building) and an energy use method (which allows completely free design of the building, provided that the overall energy consumption of the building is not increased) which may permit greater rooflight areas. These methods are rarely adopted by roofing contractors, because they require detailed knowledge about all the materials used in the building envelope, they may however be used by architects and designers at an early stage of the contract.

6.2.5.3 Control of condensation

Approved Document F; 1995 refers to BRE 262: 1994, to define measures which should be taken to prevent condensation. Built-up roofs generally require a vapour control layer (VCL) in order to prevent moisture from inside the building entering the roof cavity.

With built-up systems of lining panel and outer sheet, the VCL is best achieved by sealing all joints between liner panels, around rooflights by use of aluminium foil tape (see fasteners and sealants) and the rib voids can be ventilated by use of ventilated fillers at eaves and ridge. Factory assembled rooflights used in this type of roof structure are very difficult to seal against adjacent liners.

Where the roof is constructed using composite panels, factory assembled rooflights are the most appropriate rooflight solution.

These measures will minimize the risk of condensation. Some misting of rooflights may occur in new buildings but these measures will ensure that this should dissipate without harmful effect.

Where rooflights are required in buildings with high levels of humidity, mechanical ventilation of the building is needed to reduce risk of condensation. Sealing of liner panels and ventilation of rooflight voids is essential.

6.2.6 Loading – general

The mechanical properties of rooflights differ from those of metal and fibre-cement. They are more flexible which allows them to deflect to a greater extent without damage. Consideration must also be given to the dangers to which anyone on a such a roof is exposed, i.e. careless foot-traffic and/or access for maintenance – see section 6.2.7.3.

Standard weight rooflights have a relatively low 'pull through' performance (i.e. the force required to pull the rooflight over their fasteners). The critical conditions, for design purposes, are usually those imposed by wind suction. Deflections, with risk of 'ponding' and opening of laps, are usually only a problem on low roof slopes (6°).

6.2.6.1 Loading – wind

Wind gusts in the UK can vary from 37 m/s (over 80 mph) in the London area, to 54 m/s (over 120 mph) in the west coast of Scotland. Other factors such as topography, ground roughness, height, building shapes and permeabilities must also be considered when calculating wind loadings on structures.

As loadings on every building are different, it is recommended that wind load calculations on roofs and claddings should be completed on all major projects, on buildings sited in coastal or other exposed areas or where very large openings occur in the structure.

Wind loadings should be based on Code of Practice either CP3 Chapter V or BS 6399: Part 2; 1997, see Chapter 1.

Where necessary, rooflight manufacturers can usually provide guidance for calculating wind loads for typical buildings covered by the above Code or Standard. Load calculations on buildings outside the scope of the above document should be completed by the designer, or a structural engineer.

The pull-through performance of the fastener used in conjunction with the rooflight assembly should be determined in accordance with Annex B BS 5427: Part 1: 1996.

There are no British Standards to determine the stiffness of plastic rooflights or methods to calculate pull-through performances, these must be established by physical testing and reference should be made to the appropriate rooflight manufacturer.

The following are typical design requirements.

- Fastener washers of at least 29 mm diameter, and adequate thickness, are needed to achieve required 'pull-over' loads.
- Windload deflection should be limited to 1/15th span, or 100 mm, to prevent excessive local wear and tear around fasteners.

- Snowloads should not cause deflection of more than 1/30th span, or 50 mm, to avoid opening up sealant laps ('birdmouthing').
- Deflection on shallow pitched roofs should be limited, to avoid 'ponding'. Thicker rooflights may be needed.

The two main factors which limit spanning capability are deflection and pull-through over the fasteners. Deflection under load depends mainly on the number and depth of corrugations; pull-through loads depend on the thickness of the sheet and the number of fasteners.

The maximum span capability of a sheet under given wind loads is therefore a function of the profile (both the number and depth of the corrugations), the thickness of the sheet and the number and type of fasteners at each purlin.

As a general guide, for buildings which are less than 10 m tall; are not in an area with very high wind speeds; are not in a particularly exposed position; and have no rooflights in zones of high local windloading, where the rooflights are standard weight (i.e. GRP of 1.83 kg/m^2), have a typical profile (e.g. corrugations 32 mm deep at max pitch of 200 mm), and are fixed using a large (29 mm min. diameter) washer in each corrugation at each purlin, then the rooflights will span up to 1.8 m. [Heavier weight, GRP rooflights (2.44 kg/m^2) can span up to 2.0 m in the same situation.]

Greater spans will usually require thicker sheet. Profiles which are more rigid (due to having deeper or greater number of corrugations) will deflect less, but will not provide any greater resistance to pull over at fixings. Adding more fixings to most profiles will increase resistance to pull-over failure at fixings, but will not restrict deflection.

On tall buildings, buildings in exposed areas (e.g. by the coast or on hills) or in areas with high wind speeds, or where rooflights are located in zones of high windloads, e.g. near verges, eaves or ridge (this high loading zone has a width equal to the lesser of either the height of the building or 15% of the width), then maximum span capability may be reduced or the required weight of the rooflight increased.

6.2.6.2   Liner panels

Liner panels should not be exposed to significant wind loads. The NFRC's publication *Roofing and Cladding in Windy Conditions – 1997 edition*, gives advice on limiting wind conditions for roofwork. Fixing in windy weather should be avoided, and liners should not be left exposed before the outer sheet is fitted. Liner panels must always be secured working progressively from one end rather than securing each end before fixing at intermediate supports. Liner panel sidelaps must be secured and sealed to adjacent sheets (e.g. with aluminium foil tape), preferably with the rooflight lapping over the metal sheet on both sides.

Providing that these guidelines are followed, a liner panel with 19 mm corrugations at a maximum pitch of 250 mm will span up to 1.8 m in a ridge to eaves or chequer board configuration, and up to 1.6 m in a continuous run, as a standard weight (i.e. GRP of 1.83 kg/m^2) sheet.

If the profile is less rigid (smaller corrugations, or pitch greater than 250 mm) or the span is greater than that recommended, then some visible deflection may occur. This will only be a visual effect with no detriment to the rooflight or roof, provided the side laps are correctly sealed. Use of heavier weight liner panels is unlikely to prevent deflection: a more rigid profile or shorter span would be necessary to eliminate it.

6.2.6.3   Loading – snow

Snow loadings can be determined in accordance with procedures detailed in BS 6399: Part 3: 1988.

Provided rooflights do not exceed purlin spans indicated in section 6.6.l, rooflights in the general roof area will adequately support the snow loadings which are likely to occur in the UK.

However, rooflights should not be located in zones where exceptionally high loadings may occur due to drifts, adjacent to abutments, higher buildings, valley bottom, parapet walls and other obstructions.

CDM requirements may require heavier weight sheeting than the above wind-load figures.

6.2.7   Safety and CDM

The CDM Regulations, which came into force in March 1995, allocate clear responsibility for safety between client, designer, planning supervisor and contractor. They refer specifically to 'fragile' rooflights as an example of a potential hazard to be considered and avoided as far as possible. There is therefore a clear duty on clients, designers and roofing contractors to specify and fit 'non-fragile' rooflights.

6.2.7.1   Non-fragile rooflight tests

The only test currently available in October 1998 for establishing whether a rooflight is fragile or not is described in an HSE Specialist Inspectors Report No.30 (SIR30). This involves dropping a 45 kg 300 mm diameter bag from a height of 1200 mm on to the centre of a fully-fixed rooflight. The rooflight may be damaged, but if it prevents the bag passing through it is treated as 'non-fragile'.

The HSE has work in hand (in 1998/9) to develop a better test; until this is complete, SIR30 is the only published test. Many rooflight manufacturers are independently carrying out tests, to enable them to advise specifiers, clients and contractors how to achieve 'non-fragile' rooflight installations.

6.2.7.2 Test results

Most rooflights are flexible, and deflect under the drop test. They may be damaged. Resistance to penetration depends almost entirely on the fixings; number, type and method of fixing are critical to achieving non-fragility.

All such flexible rooflights should be treated as 'fragile', and precautions taken as detailed in HSG 33, *Health and Safety in Roof Work*, until they are fully installed and fixed in accordance with specific manufacturers recommendations. Paragraph 3.1 of this document gives typical recommendations, arising from tests carried out by the manufacturers. Properly fixed, most GRP and polycarbonate double-skin rooflights pass the above test.

It is recommended that consideration is given to extra safety reinforcement when using PVC. However, even 'non-fragile' rooflights are likely to be damaged by impact, and crawling boards must be used at all times, except with heavy weight rigid rooflights designed to withstand foot traffic.

6.2.7.3 High risk areas

It is recommended that rooflights should not be located at eaves adjacent to valley or boundary wall gutters which can be used as walkways, or adjacent to roof services, unless they are designed to resist impact without damage.

6.2.7.4 Maintenance access

All rooflights should be considered 'fragile', unless the rooflights are clearly marked as Safe, or unless a Health and Safety File is available which confirms the non-fragile nature of the rooflight and their installation, and their anticipated safe life.

Although on newer buildings, rooflight assemblies may have been designed and installed to withstand impacts from falling bodies or inadvertent foot traffic, flexible rooflights are likely to be damaged and will deflect under point loadings which may damage sealant or fixings. Crawling boards should be used, and any rooflight which suffers damage which is likely to reduce its strength must be replaced.

To help alert maintenance personnel to the presence of rooflights, primary and secondary fasteners should be brightly coloured, providing a contrast with the roof sheeting. Coloured fastener heads should be factory coloured, and not loose push-on caps which can easily come off. Experience has shown that poppy red is a suitable colour on most roofs.

6.3 SITE ASPECTS

6.3.1 Fasteners

The performance and durability of flexible rooflight is heavily dependent on the fasteners, and the integrity of installation. The number and position of fasteners is also critical to achieving 'non-fragile' performance,

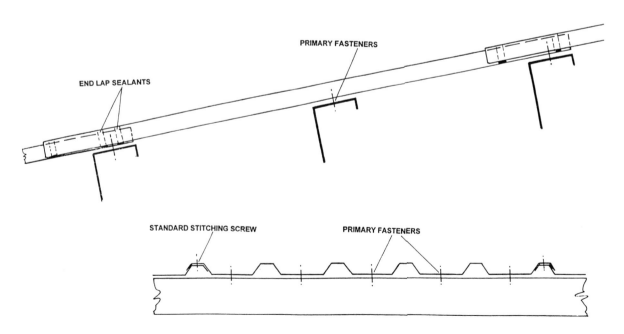

Figure 6.1 Single-skin rooflight.

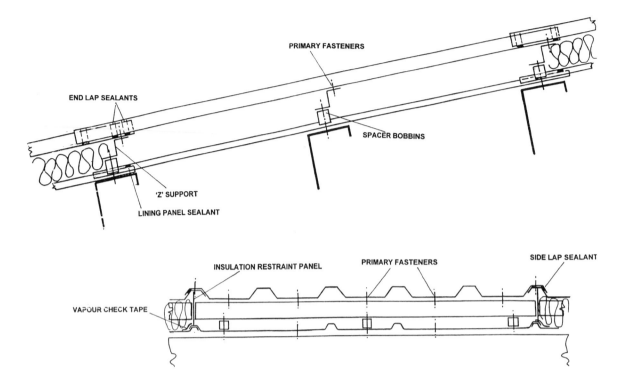

Figure 6.2 Double-skin site assembled rooflight.

i.e. ability to withstand drop tests. The requirements for the safe application of rooflights, in respect of the reduction or elimination of risks as deemed necessary by the CDM regulations are contained in Health and Safety Guidance Note HSG 33. The following fixing recommendations are typical for GRP rooflights, to BS 4154. If in doubt, manufacturers should be consulted.

They are designed to ensure that, for roof slopes down to 6°:

- liner panels are sealed to minimize condensation risk;
- weather sheets resist wind-uplift, snow loads and water penetration.

Note: Special precautions are needed for roof slopes below 6°.

Liner panels

Maximum sheet length	– 2 purlin spans of 1.8 m
Minimum sheet weight	– 1.83 kg/m²
Number of fasteners	– at least 5 per metre at each purlin (including ferrules/ brackets)
Type of fastener	– minimum 5.5 mm diameter and 29 mm washers
	– fasteners at least 50 mm from end of sheet, to allow for

tearing of sheet under heavy impact.

Laps	– end laps sealed with mastic.
	– side laps – GRP overlap metal both sides.
	– sealed with foil tapes applied over laps.

Notes:
- Sheets should be fixed progressively, from one-end.
- Liner panels fixed in windy weather or left exposed to winds can become stressed, and lead to visible deflections in service.

Weather sheets

Minimum sheet weight	– 1.83 kg/m² for double-skin application
	– 2.44 kg/m² for single-skin application (heavier weights will be needed for spans over 1.8 m)
Primary fasteners	– 5 per metre (200 mm pitch maximum).
	– Minimum 5.5 mm diameter, 29 mm washers.
	– Fasteners at least 50 mm from end of sheet.

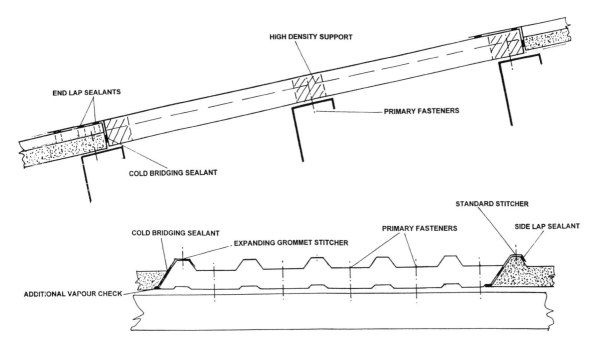

Figure 6.3 Double-skin factory assembled rooflight.

– For metal profiles with trough widths less than 40 mm, use smaller washers and increase number of fasteners.
– Avoid crown fixings, except with manufacturer's advice, or when fixing to fibre-cement sheeting.
– Use brightly coloured factory applied integral caps on fasteners.
– Locate fasteners in centre of top flange of Z-spacer (or equivalent support).

Secondary fasteners (side laps)
– 400 mm maximum centres, 300 mm in exposed areas
– Rooflight to overlap both sides whenever possible
– Standard (minimum 4.8 mm) stitching fasteners into metal, or expanding grommets where GRP underlaps
– Specify purlin bearing leg where GRP underlaps
– Use brightly coloured factory applied integral caps on fasteners

Notes:
• All washers should incorporate bonded EPDM seal.
• Avoid light-weight washers, and avoid flattening washers by over-tightening.

• Fix sheeting progressively down roof – do not fix ends first.
• Fasteners must be square to roof.
• Pre-drill oversize concentric holes, for PVC and polycarbonate (see section 6.3.3)
• Fasteners should be installed without over-tightening so that the washers are correctly compressed.
• With fibre-cement profiles, use crown fixings with fillers to support profile.

6.3.2 Sealants

For weather sheets, cross-linked butyl mastic, white or light in colour, is preferred. Pre-formed strips or beads are more consistent and safer than gun-applied sealant.

Position is very important to ensure weather tightness. End lap sealants should be placed immediately either side of the primary fixing centres (max. 15 mm from the centre line) and be laid well into the troughs, to allow washers to compress the sealant and keep the end lap true to roof line. Where rooflight overlaps rooflight on end laps or PVC rooflight overlaps metal, a further seal of clear silicone close to the end of the lap will prevent dirt entering the lap. This should be totally sealed to prevent water entering.

Weather sheet side laps should be sealed with at least one strip of sealant, outboard from the fixing line.

Lining panels should, where possible, overlap metal on either side and be sealed with adhesive foil or tape applied over both lapping sheets. Liner panel's end laps can be sealed with either adhesive foil/tape, or a single strip of butyl mastic along the line of the fixings.

6.3.3 Thermal expansion

All materials used in the construction of a building expand and contract to temperature change.

Typical expansion coefficients for roofing materials are:

Aluminium	23×10^{-6}
Mild steel	12×10^{-6}
GRP	22×10^{-6}
PVC	68×10^{-6}
Polycarbonate	65×10^{-6}

PVC and polycarbonate sheet require oversize holes to allow the sheet to move without putting stress on it from the fixing. Oversize holes must be at least 4 mm larger than fixing diameter for sheet length of 3 m and 6 mm larger for sheet length of 4 m. A sheet longer than 4 m should not be fitted due to the excessive thermal movement in its length. The fixing must always be fitted in the centre of the oversize holes, this can be achieved by pilot drilling prior to drilling the oversize hole. It is recommended that stepped drill bits are used to ensure the correct size holes.

GRP rooflights do not normally require any provision for thermal movement. Careful procedure is needed for roofs with expansion joints, or liner panels with minimum ribs.

6.3.4 Fixing sequence

Rooflights should always be fixed progressively from one end, and not be fixed at both ends before fixing at intermediate purlins, as this can cause stresses to be trapped.

Whenever possible, avoid fixing rooflights in windy weather, or leaving liner panels exposed before fitting outer sheets. Exposed liner panels can become stressed during high winds, and subsequent thermal movement may lead to visible deflection.

Thin purlins can be distorted if loaded with heavy stacks of sheet, and may result in flexible rooflights deflecting.

6.4. TRANSPORT, HANDLING AND STORAGE

All rooflights should be handled and stacked with care. Surfaces are easily scratched, and heavy stacks can damage lower sheets.

All rooflights should be stored flat, the right way up, on 75 mm wide battens not more than about 1.8 m apart, free from nails. Stacks should not be higher than 1 m, and must be covered and protected from rain and sun. Exposed stacks can permanently discolour, due to action of sun and water. Thermoplastic sheets can overheat and deform in a stack. Refer to manufacturer's recommendations.

Different profiles in a stack need battens between profiles – but battens must be vertically above each other to avoid over-loading. Factory-assembled units require very careful handling and storage.

6.5 MAINTENANCE

First maintenance (typically 10 years, depending on conditions) should involve cleaning down with warm water and stiff bristle brush and cleaning off any grease or tar deposits with white spirit. When clean and dry, paint with acrylic or polyester lacquer to refurbish the sheet surface.

The only other maintenance required is inspection for damage, and check on fixings for tightness (without overtightening), and condition of sealants, every 2–5 years, depending on environment.

Rooflights should not be painted over with an opaque covering. This can be dangerous, and may cause premature failure.

During maintenance access, all rooflights should be treated as fragile, and full precautions observed, unless they are clearly and all unmistakeably marked as 'Safe', or unless a Health and Safety File is available which clearly indicates that the rooflights have an adequately long 'non-fragile' life. To prevent possible damage to fixings and sealants, the use of crawling boards is recommended at all times.

7

FASTENERS AND FIXINGS

7.1 INTRODUCTION TO FASTENERS AND FIXINGS

The term 'fastener' is used to describe a mechanical device used to secure any cladding component to a structure or to another component. A 'fixing' describes the result achieved by the use of a fastener.

Thus the strength of a fastener denotes certain mechanical properties of the item in isolation, whereas the strength of a fixing must take account of the 'pull-out', 'pull-over', 'washer inversion' and 'over-torque' values, which reflect the properties of the member into which the fastener is secured and the sheet or component being restrained.

7.1.1 Primary fasteners

Primary fasteners are the fasteners which are relied upon for structural performance. Thus fasteners used to secure profiled sheet metal to either purlins or spacers and fasteners attaching the spacers to purlins are all load-carrying or 'primary' fasteners and demand to be considered accordingly (see Figure 7.1).

7.1.2 Secondary fasteners

Secondary fasteners are not generally relied upon to contribute to structural performance but may be involved in transferring loads from sheet to sheet.

Secondary fasteners are used to stitch side laps, flashings, etc.

It follows from the definitions that primary fasteners must be designed so that they safely carry the relevant loads determined in accordance with the Building Regulations (or other relevant legislation) and must be rendered weathertight where the fastener head is exposed to the weather.

Secondary fasteners cannot be dismissed as non-load bearing and must be rendered weathertight where the fastener head is exposed to the weather.

7.2 MODES OF FAILURE

7.2.1 Principal modes of failure of fasteners in service

These are as follows.

1. Pull-over   – where the service load causes the sheet to be pulled over the fastener head due to a combination of incorrect washer diameter and/or insufficient number of fasteners.

2. Pull-out   – where the fastener is pulled out of the support member, due to insufficient number of fasteners and/or incorrect choice of type or failure to install correctly.

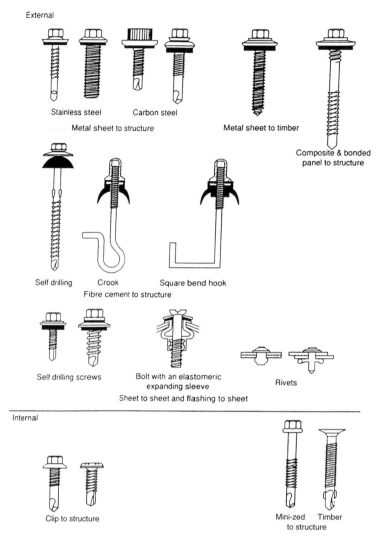

Figure 7.1 Fastener types (reproduced with permission of BSI).

3. Torque-strip – where the fastener is over driven and strips the threads, or the pre-drilled hole if any had too large a diameter.

4. Back-out – where the fastener works loose, often due to the incorrect selection of thread type for the thickness of materials.

5. Shear – may be found where the fastener diameter has been reduced by corrosion, or differential movement between sheet and support.

6. Corrosion – often resulting from incorrect selection or application of the fastener's material for the type of environment.

7.2.2 Standards and test methods

BS 5427: 1996 gives basic methods in Annex B for 'Determination of the strength of attachment of a cladding system'. A range of recognized test methods have been adopted by fastener manufacturers, which may not produce directly comparable results. The roofing contractor should obtain performance values from the fastener supplier before selecting the fastener type and frequency, to ensure they exceed the design loads for the intended use. The values considered should be for the appropriate type and thickness of both the profiled sheet and the intended support member which may be the spacer bar.

7.3 TYPES OF FASTENERS

7.3.1 Primary fasteners

7.3.1.1 Self-drilling self-tapping fasteners
This type of fastener is often preferred by the contractor due to their speed of installation. They typically have a

minimum thread diameter of 5.5 mm when used as a primary fastener.

A range of drill-point lengths and diameters are available designed to give optimum pull-out and strip-out performance in various specific thicknesses of purlin or spacer.

To determine the correct thread length, and thus the fastener length, the thickness of each material in the build-up must be added together.

Where there are additional materials between the sheet and spacer (e.g. compressed insulation on wall cladding applications), fasteners with 'pilot points' are available to prevent jacking and provide a good clamping action.

7.3.1.2 Thread-forming self-tapping fasteners

This type of fastener is typically of at least 6.3 mm thread diameter for use as primary fasteners and are generally of two basic types:

1. gimlet pointed, e.g. Type A, AB, and C;
2. blunt ended, usually with a slight taper or lead-in, to facilitate initial engagement into hot rolled steel, e.g. Type B.

These fasteners require pre-drilled pilot holes and function by thread forming (not thread cutting), producing their own mating threads by forcing the metal to flow round the screw threads. The gimlet pointed types are usually limited to use in cold rolled steel purlins and spacers of stated minimum to maximum thickness range (and are also used for fixings into timber).

The blunt ended B type is more likely to be given a stated minimum material thickness and be approved for use into hot rolled steel sections.

It is essential to use a drill of the diameter specified by the fastener manufacturer appropriate to the type and thickness of the material into which the fixing is to be made. The use of worn drill bits should be avoided as this will lead to inconsistent and/or incorrect hole geometry.

An oversized hole will give a fixing of reduced pull-out performance, whilst an undersized hole is likely to result in failure to achieve penetration, or it will produce an increase in torque which may cause overstraining or breakage of the fastener.

7.3.1.3 Thread-cutting self-piercing fasteners

These are available as 'S' pointed and are normally limited in use to fastening into timber. They are typically at least 6.3 mm thread diameter. These differ generally from thread-forming fasteners by having a flute at the tip to provide cutting edges on the first few threads.

7.3.1.4 Insulated composite panels

Special designs of fastener are used with both site and factory assembled insulated composite sandwich panels. For fixing through the composite panel on roofs these special fasteners are dual threaded.

A normal thread is used to fix into the support member with a second larger diameter thread under the washer head. This is intended to support the outer skin of the system ensuring long term weathertightness without over compression of the insulation. The in service performance of these support threads varies, fasteners for composite panels should therefore only be selected according to the composite panel manufacturer's recommendations.

7.3.1.5 Secret-fix systems

Specialized primary fasteners, usually a form of clip and bracket, are supplied to suit specific systems. The fastener which holds the clip or bracket should only be selected according to the systems manufacturer's recommendations.

7.3.2 Secondary fasteners

7.3.2.1 Self-drilling self-tapping fasteners

The description and comments given above for primary fasteners are equally applicable to secondary fasteners. Purpose designed secondary (stitcher) fasteners are available with special features to provide high clamping forces when used to join thin sheeting.

There are two types for secondary fastening purposes, the first having the reduced drill point and a parallel, self-tapping thread with a spin free zone under the head. The alternative has a tapered thread, which is often described as a 'high torque stitcher'.

7.3.2.2 Rivet type fasteners

The final group of secondary fasteners comprises those, whether of metallic or non-metallic construction which employ a riveting action. The most widely used are of aluminium alloy.

Caution: On fire wall construction only steel rivets or fasteners should be used, at the spacing given in the profile manufacturer's test certificate.

The mandrel-operated expanding rivet is intended for use with lever arm tools or lazy tongs and requires a pre-drilled hole. It is recommended that this type of rivet be used with a sealing washer when used in roofing applications, unless the rivet's manufacturer has shown that the rivet is watertight without the use of a compressible washer.

As with any type of rivet, the length needs to be determined with care and must allow for each component of the lapped joint, including washer and sealant. If it is too long, a loose fixing is the probable result, too short and the mandrel is likely to break prematurely, again producing a poor joint.

Another type of rivet-action fastener is moulded in an elastomeric material (EPDM, Neoprene or similar) and derives its action from a screw engaging with a 'nut' moulded into the body of the fastener. The force exerted by the fastening action is cushioned by the elastomer and for this reason the fastener is often recommended for use with rooflights of plastics material. The screw should be manufactured from austenitic stainless steel, as a protective cap cannot be fitted on some designs.

The resistance of the fastener material to UV radiation should be checked with the manufacturer before use in sunlight.

7.3.3   Other fixing applications

In addition to the fastener types mentioned above, there is a wide range of further fastener (fixing) solutions available for other applications, e.g. fastening into concrete/masonry, securing timber to steel. Consult the fastener supplier (e.g. an NFRC Associate Member) to determine the appropriate solution to the fixing detail.

7.3.3.1   Liner panel to structure fasteners
In site assembled liner panel systems it is common practice to pre-fix the non-structural liner panels with a fastener similar to that described in section 7.3.1.1 of minimum 5.5 mm diameter with washers.

The impact force of a falling body on a thin metal liner sheet may tear the sheet away from its fasteners, therefore a minimum of three fixings per liner panel at each purlin are recommended for safety. Under current HSE and CDM regulations, liner sheets should be deemed as fragile unless the supplier has advised that independent tests have proved the liner as non-fragile in accordance with the current regulations.

Where a metal liner sheet is used as the vapour control layer the screw fasteners with washers through the lining sheet must provide the highest possible standards of vapour seal.

7.3.3.2   Rooflight liner to structure fasteners
To reduce the risk of falling through a rooflight, a minimum of five fixings of minium 5.5 mm diameter with 29 mm washers per rooflight liner panel at each purlin may be required for safety. These fasteners should be at least 50 mm from the end of the sheet,

see Chapter 6 for safety precautions when fixing rooflights.

7.3.3.3   Hook bolts and crook bolts
Traditionally 8 mm diameter J bolts, hook bolts, crook bolts and square bent hook bolts provide a method for fixing corrugated (sinusoidal) steel and non-metallic sheet to hot-rolled steel angle or precast concrete purlins and rails. Proprietary nuts and washers which have adequate corrosion resistance and shape which will seal to the profile crown should be used. Special safety precautions are required when using these types of fasteners and contractors are urged to refer to the various Health and Safety and CDM regulations, see Chapter 10.

When using this type of fastener, it is important to ensure that contact between sheeting rails and fastener takes place at the seat of the hook. This is essential to achieving the full strength of the fixing.

7.4   FASTENER MATERIALS

7.4.1   Choice of materials

Fasteners are not in general replaceable or maintainable and should therefore have a functional life equal to that of the roof or wall cladding which may range from under 10 years to over 25 years. Certain fastener materials will have reduced durability in severe environments or when in contact with some types of materials.

Guidance on adverse environments can be found in BS 5427: 1996 and design life in BS 7543: 1992.

Coated carbon steel would be considered suitable for the majority of industrial and commercial buildings where there is generally not a risk of high internal humidity or corrosive atmospheres and where a functional life of up to 20 to 25 years is required. Protective coatings on carbon steel are necessary to extend the fastener life.

Austenitic grades (typically 304 to BS 1449: Part 2 or BS 970: Part 1) of stainless steel make, in general, more durable fasteners than coated carbon steel and are available in most fastener types as noted in section 7.3 above. Care is needed when installing, as the softer threads are more vulnerable to stripping than carbon steel. It is, therefore, essential that the manufacturer's recommended maximum drilling speed or torque setting are not exceeded.

Austenitic stainless steel should be considered for buildings with high humidities, polluted industrial atmospheres, marine environments, those where hygiene is critical and where an extended life over that offered by coated carbon steel is required. Stainless steel is recommended for use in fixing aluminium

Table 7.1 Guide to selection of exposed fastener materials

Fastener material	Environment		Functional life expectancy (years)	Sheet material (see Note 1)			
	Internal humidity grade	External exposure		Aluminium	Coated steel	Stainless steel	GRP/PVC fibre-cement
Minimum† washer dia.				19 mm	19 mm	19 mm	29 mm
Coated carbon steel and push-on plastic caps	Dry/low humidity 'A and B'	Urban/Rural	10/20	NR	✓	✗	✓
		Industrial*	10/15	NR	C	✗	✓
		Coastal/Marine	10	✗	C	✗	C
	High humidity 'C'	Urban/Rural	10/15	NR	C	✗	C
		Industrial*	10	NR	C	✗	C
		Coastal/Marine		✗	✗	✗	✗
Coated carbon steel with integral plastic coloured head	Dry/low humidity 'A and B'	Urban/Rural	15/25	NR	✓	✗	✓
		Industrial*	15/25	NR	✓	✗	✓
		Coastal/Marine	10/15	NR	C	✗	C
	High humidity 'C'	Urban/Rural	10/15	NR	✓	✗	✓
		Industrial*	10	NR	C	✗	C
		Coastal/Marine	10	✗	C	✗	C
Austenitic stainless steel	All humidity grades	Urban/Rural	25 +	✓	✓	✓	✓
		Industrial*	25 +	✓	✓	✓	✓
		Coastal/Marine	20	✓	✓	✓	✓
Aluminium secondary fasteners, e.g. rivets	All humidity grades	Urban/Rural	20/25	✓	✓	✗	C
		Industrial*	15/20	✓	✓	✗	C
		Coastal/Marine	15/20	C	C	✗	C

✓   Recommended for use in these environments.
✗   Unsuitable for use in these conditions.
C   Check suitability of both sheet system and fastener with manufacturer.
NR   Not recommended for use with aluminium sheets by some profile manufacturers.
*   Subject to non-polluted environment, may not be suitable in corrosive or other chemical laden conditions.
†   Minimum washer diameter is quoted for roofs, smaller diameter may be suitable for walls.
Note 1   This table gives guidance on the selection and functional life of the fastener in various sheet materials. Consult the sheet manufacturer regarding the most appropriate sheet material and coating and its functional life in the particular environment.
Note 2   Fasteners exposed only to internal, non-aggressive environments would have a longer functional life than stated in the table.

sheeting where the failure of any barrier coating on carbon steel could lead to bimetallic corrosion.

The most appropriate choice of fastener for any particular project will be selected as a result of detailed consideration of risk associated with the internal and external environment, as well as the recommendations of the particular system supplier. The selection of fasteners is important, incorrect choice can cause premature failure of the roof or wall cladding, general guidance on suitability is given in Table 7.1.

The humidity grades of environment found within a building can be classified as follows.

Grade A – Normal humidity

Factories and warehouses for normal manufacturing and storage purposes where the occupants or processes do not add significant quantities of water vapour to the atmosphere.

Grade B – Medium humidity

Buildings where large numbers of people congregate, eg. public meeting halls, supermarkets, offices.

Buildings where people meet for exercise, or where heating is used only for short periods, e.g. sports halls, youth clubs, church halls.

Grade C – High humidity

Swimming pools, or buildings containing wet processes.

Full details of humidity classification are given in Chapter 2. These environmental classifications are used in Table 7.1 which is a guide to external exposed fastener material selection

7.4.2 Fastener heads

It is important to use the correct drive socket and screw guns fitted with a correctly set depth locator to avoid head damage, especially when using integral colour headed fasteners. There are three sizes of head in common use and each requires a matching socket and depth locator if the washer is to be correctly compressed.

Countersunk fasteners and rivets require matching tool shapes and the manufacturer should provide the appropriate information.

7.4.3 Exposed fasteners – colour matching

Push-fit plastic caps provide one method for colour matching purposes and, if correctly designed and applied, may give added protection from the elements. Plastic caps must be of a material which is resistant to the weather and the UV rays of the sun.

Factory produced integral colour heads provide a longer term and more reliable solution to colour matching and also provide additional corrosion protection to the fastener head from the external environment. However integral colour heads should not be relied upon to provide the sole means of corrosion protection to the fastener, see section 7.3.1.

Factory fitted integral colour heads should be considered for walls where access to replace loose caps which have fallen off can be costly.

As an added precaution, to help alert maintenance personnel to the presence of rooflights primary and secondary fasteners should be brightly coloured, providing a contrast. Colour fasteners should be factory coloured and not push-on caps which can easily come off Experience has shown that poppy red is a suitable colour on most roofs.

7.4.4 Fastener washers

Many types of roofing and cladding systems, other than the secret-fix type, rely upon the use of washers and resilient seals to achieve weatherproofing.

Washers vary widely in design from the original heavy gauge diamond pattern used as a standard item on galvanized and asbestos cement corrugated sheeting to the circular 'hi-tech' types with integral seals.

Washers are available in galvanized steel, aluminium alloy or stainless steel.

The range of plastics materials used for the sealing medium includes PVC (plasticized), butyl rubber, Neoprene and EPDM (ethylene–propylene–diene–monomer).

The seals must retain their elasticity under all conditions to which they are exposed in service, whether they be extremes of temperature, moisture, UV light, ozone, general environmental contaminants or oil film on sheeting surfaces.

EPDM is generally recognized as providing the best all round performance for compressible sealing washers.

Circular conical washers, designed to compress the seal around the fastener and against the sheeting, are available with various resilient materials, either bonded to or captive to the metal backing.

For the seal to be effective, the washer must not be deformed excessively whilst exerting pressure uniformly over the surface of the sealing material which must exclude moisture from under the fastener head and from the sheeting interface without extruding significantly beyond the washer and interfering with the fit of any plastic cap (see Figure 7.2).

Another important function of the washer and fastener head design is to reduce the risk of failure by 'pull-over' when due to uplift forces acting on the roof, the weather sheet is 'pulled-over' the fastener.

Circular washers vary in diameter from nominally 13 mm to 32 mm, the tendency being to use the smallest on vertical cladding and the largest on plastic rooflights. For valley fixings, the washer size may be limited by profile dimension but, in general, the use of intermediate sizes will either be on sheeting manufacturer's recommendation or as a result of experience with a particular type.

For general guidance, the minimum diameter of washers for primary fasteners should be as Table 7.2, unless specified otherwise by the sheeting manufacturer.

All types of washer are dependant upon care being exercised by the installer not to under or over tighten the fastener. It is most strongly recommended by all fastener manufacturers that electric screwdrivers are fitted with correctly set depth locator devices or torque

Figure 7.2 Washer sealing pressure.

Table 7.2 Guidance for the minimum diameter of washers for primary fasteners

Profiled sheet material	Roofing applications minimum diameter (mm)	Walling applications minimum diameter (mm)
Aluminium	19	19
Steel	19	15
Translucent rooflights	29	

control, to ensure correct and consistent compression of the sealing washer.

Preference for securing trapezoidal profiles is in their trough, crown fixing may be considered for composite panels and special systems. Saddle type washers are available designed specifically for crown fixing. This provides an extended surface area in contact with the sheeting to minimize the risk of profile distortion under fixing pressure. The saddle washer has its own resilient gasket but is recommended for use with a normal size sealing washer beneath the fastener head. The fastener used through a saddle washer must be centred on the crown of the sheet profile, otherwise there is a risk that the saddle washer will not be seated correctly and may leak.

7.5 FASTENER APPLICATION

7.5.1 Forces acting on fasteners

Chapter 1 indicates the nature of the loads acting upon the cladding, which relate directly to the performance required of the fasteners.

It is recognized that the wind forces and internal pressure coefficients are responsible for the net uplift on roof sheets and that they do not act uniformly across the entire roof area, thus different areas of sheeting will be subject to varying upward forces.

Because these forces must be carried by the fasteners, it follows that, theoretically, the distribution of fasteners will not necessarily be uniform over the entire area. Attention is drawn to the reference in Chapter 1 concerning the abnormal loads likely to be carried by structures in course of erection.

The fastener selection will have regard for all relevant modes of stress (tensile, shear and bending), the effects of thermal expansion, potential for corrosion and location of the fixings (whether in the profile trough or crown).

The sheeting and fastener manufacturers respectively will provide, on request, data supported by test results, for pull-over, pull-out, washer inversion and torque-strip failure to enable the designer to apply the relevant safety factors and, with the consideration of all loading requirements, determine the fastener type and frequency.

The designer must also recognize the fact that the thinner the gauge of structural member the greater the need for close control of fastening technique because of the reduced margin for error.

7.5.2 Designer's responsibilities

The designer bears the responsibility for assessing the loadings and, aided by the advice of the sheeting and fastener manufacturers, for specifying the type of fastener to be used. The following items will need to be considered in making this specification:

- fastener type;
- diameter, length and thread pattern to suit sheet and spacer or purlin thickness;
- material durability and compatibility;
- washer diameter, thickness and material;
- distribution of fasteners over the cladding;
- determining spacer dimensions to suit the design loads and thermal insulation thickness.

If the fastener supplier or manufacturer does not make satisfactory product test data available, the designer should select fasteners from an alternative supplier.

If timber spacers are to be used, care must be taken in the choice of preservative treatment, to avoid corrosion of the fastener material. The spacer thickness should be sufficient to allow two-stage fastening.

7.5.3 Sheeting contractor's responsibilities

The sheeting contractor is responsible for complying with the specification and executing the work in a competent manner as befits a specialist.

The following are some of the basic principles to be observed in relation to general fixing procedures.

1. To ensure that the site operatives have received appropriate training (including safety training).
2. To ensure that the site operatives have the necessary drawings and information, tools and equipment, in fit condition, to properly carry out the work.
3. To set out cladding and fixings as recommended by the supplier of the lining system; sheeting or

concealed-fix panels; composite or semi-composite system, as applicable.

4. To ensure that fixings are made with due regard to:

   (a) the correct location of primary fasteners (crown or valley) according to sheeting manufacturer's recommendations.

   (b) the use of the correct fastener at each fixing including sheet-support/spacer types as required, and the appropriate types of sealing washer (e.g. circular or saddle) to sheeting manufacturer's recommendations;

   (c) the correct positioning and spacing of secondary fasteners and use of approved sealing washers;

   (d) the correct sized pilot holes for non-self-drilling fasteners;

   (e) the need for correct sized holes to be drilled (punched holes should only be used if recommended by the material supplier) perpendicular to the surface of the sheet and for drilling swarf to be carefully removed;

   (f) use of correct drill speeds (standard twist drills and self-drilling fasteners alike);

   (g) proper maintenance of drills and electric screwdrivers for optimum results;

   (h) the need to avoid over-tightening fasteners with adverse effect on washers and seals or thread-stripping (especially in light gauge materials); always use power tools with either correctly adjusted depth locators or torque control devices;

   (i) the need to maintain correct end and side laps;

   (j) the correct use and application of lap sealants;

   (k) the correct alignment of spacers to ensure sound fixings whether into metal or timber and the use of two-stage timber fixings as standard procedure;

   (l) the need to avoid compressing foamed insulation causing permanent deformation, whilst ensuring adequate pressure for an effective seal;

  (m) the requirement to cut and fit mineral wool insulation tightly between spacers;

   (n) the use of fixings, sealants and lap arrangements for rooflights in strict accordance with the manufacturer's recommendations – if none are supplied, ask.

8

SEALANTS AND PROFILE FILLERS

8.1 INTRODUCTION

8.1.1 Reasons for use of sealants

At shallow pitches the side and end laps of profiled sheeting require effective seals as a means of achieving a weathertight building envelope. Recommendations for the position and application of sealants are given in the relevant roof material chapters, the recommended sealant types are given in the following corresponding tables:

1. profiled metal weathersheet and composite panels, (Chapters 3 and 4), Tables 8.1–8.3;
2. fibre-cement weathersheets (Chapter 5), Table 8.4.

Sealants are also essential to the formation of effective vapour control layers (VCL) necessary for limiting the risk of interstitial condensation.

8.1.2 Profile fillers

With profiled roof and wall cladding sheets it is essential that the troughs are filled at intersections, e.g. ridge trims. Profiled foam fillers are widely used for this purpose with trapezoidal metal and translucent sheeting.

8.2 FUNCTIONS AND DESIRABLE PROPERTIES OF SEALANTS

8.2.1 Functions of sealants

In the context of this Guide, sealants have two different functions:

1. to prevent the entry of rain and snow through lapped joints between profiled sheets having a variety of surface treatments: between those sheets and plastics materials and between overlapping sheets of plastics material –
   (a) when the overlap is fixed, and
   (b) when the overlap constitutes a movement joint designed to allow relative movement between the sheets;
2. to prevent the passage of moisture vapour from within the building through lapped joints in various lining materials, under the influence of vapour pressure differential.

A requirement common to both duties is that the chosen sealant must remain effective throughout the design life of the roof.

There is no reason to suppose that one sealant will satisfy these differing requirements and the selection of suitable compounds is a matter for advice from the

Table 8.1 Profiled steel sheets

Sealant type	Life (years)	Profiled steel sheets			Metal liner	
		Side lap	End lap	Flashing	Side lap	End lap
Preformed sections						
PIB butyl	20	Y	Y	Y	Y	Y
Cross-linked butyl	20	Y	Y	Y	Y	Y
Self-adhesive tapes						
EPDM foam	20	C	C	N	C	C
PVC foam	15	C	N	N	C	C
Polyethylene foam	15	C	N	N	C	C
Aluminium backed		N	N	N	Y	C
Gun applied sealant						
Silicone (neutral cure),						
1 part *	20	C	C	Y	C	C
Polyurethane, 1 part *	20	C	C	Y	C	C
Butyl	10	C	C	Y	C	C
Grease		N	N	N	N	N

Y Recommended.
N Not recommended.
C Check suitability with sealant manufacturer and profile manufacturer's recommendations.
* Primer may be required.

Table 8.2 Profiled aluminium sheets

Sealant type	Life (years)	Aluminium profiled sheet			Aluminium liner sheets	
		Side lap	End lap	Flashing	Side lap	End lap
Preformed sections						
PIB butyl	20	Y	Y	Y	Y	Y
Cross-linked butyl	20	Y	Y	Y	Y	Y
Self-adhesive tapes						
EPDM foam	20	C	N	N	C	C
PVC foam	15	N	N	N	C	C
Polyethylene foam	15	C	N	N	C	C
Aluminium backed		N	N	N	Y	C
Gun applied sealant						
Silicone (neutral cure),						
1 part *	20	C	C	Y	C	C
Polyurethane, 1 part *	20	C	C	Y	N	N
Butyl	10	C	C	Y	N	N
Grease		Y	Y	Y	C	C

See footnotes to Table 8.1.

specialist manufacturers. It is not a matter to be left to the discretion of the storekeeper!

8.2.2 Properties required of sealants

The properties required of ideal sealants can be summarized as follows.

8.2.2.1 As a weather sealant
The properties required are:

1.* stability in storage;
2.* ease of application under the ambient and surface temperatures encountered during construction;
3.* adhesion to the overlapping materials in the sense of 'clinging to' rather than 'restraining' the materials in movement joints;
4.* the ability to remain within the lap, resisting slump and creep;
5.* low resistance to compression;
6.* water repellency;
7.* no undue shrinkage;

Table 8.3 Miscellaneous sheets

Sealant type	Life (years)	Gutters		Rooflights		Insulation boards
		Steel	Aluminium	PVC	GRP	Board to T bar
Preformed sections						
PIB butyl	20	Y	Y	Y	Y	N
Cross-linked butyl	20	Y	Y	Y	Y	Y + tape
Self-adhesive tapes						
EPDM foam	20	C	C	C	C	N
PVC foam	15	Y	Y	C	C	N
Polyethylene foam	15	Y	Y	C	C	N
Aluminium backed	15	C	C	C	C	Y+cross-linked butyl
Gun applied sealant Silicone (neutral cure),						
1 part *	20	Y	Y	Y	Y	C
Polyurethane, 1 part *	20	Y	Y	Y	Y	N
Butyl	10	N	N	C	C	N

See footnotes to Table 8.1.

Table 8.4 Profiled fibre-cement sheets

Sealant type	Life (years)	Profiled sheet			Liner sheet	
		Side lap	End lap	Flashing	Side lap	End lap
Preformed sections						
PIB butyl	20	Y	Y	Y	Y	Y
Cross-linked butyl	20	Y	Y	Y	Y	Y
Self-adhesive tapes						
EPDM foam	20	C	C	C	C	C
PVC foam	15	C	C	C	C	C
Polyethylene foam	15	C	C	C	C	C
Aluminium backed	15	C	C	C	C	C
Gun applied sealant Silicone (neutral cure),						
1 part *	20	C	C	C	C	C
Polyurethane, 1 part *	20	C	C	C	C	C
Butyl	10	N	N	N	N	N

See footnotes to Table 8.1.

8.* no adverse reaction in contact with roofing materials and their surface finishes;

9. resistance to ageing, weathering and UV radiation;

10. ability to accommodate movement in shear without breaking down, even in thin bands of sealant;

11.* ability to be drilled and receive fasteners when under compression, without being displaced;

12. physical properties retained over a wide annual range of surface temperature – UK guide values:

light coloured sheeting – 20 to 60°C
dark coloured sheeting – 20 to 80°C
clear rooflights – 20 to 50°C

13.* no maintenance required;

14.* long life expectancy.

8.2.2.2 As a vapour sealant

* A sealant suitable for use as a component in the formation of a vapour control layer should possess the properties marked * in the above schedule of 'weather sealant' characteristics and in addition should have the following properties.

15. the permeability of the seal to moisture vapour must be consistent with the resistance required of the vapour barrier.

16. there must be no shrinkage which might impair the vapour resistance of the seal.

17. although not exposed to the elements, the sealant must be resistant to ageing and UV radiation.

18. the physical properties must be retained under

indoor conditions relevant to the usage of the building and not be adversely affected by short-term exposure to the conditions in section 8.2.2.1 above.

Sealant manufacturers should be consulted on their ability to meet the relevant specification and be invited to examine samples of the associated materials, finishes, profiles and representative joints before making their recommendations.

8.3 FORMS OF SEALANT AVAILABLE

Sealants are available in various forms:

1. Pre-formed sealant (mastic) strip which should have the backing paper attached whilst being applied to avoid placing the sealant under tension, with consequent reduction in cross-section and the risk of elastic 'memory' causing shrinkage and discontinuity.

   Soft grades are needed if joints are to be pulled together without bulging of the sheet between fasteners. The manufacturer's advice should be sought regarding the correct cross-section to ensure adequate compression for an effective seal.
2. Cellular foam strip produced from EPDM, PVC or other plastics, in a range of sections and densities which require from about 25% to 50% compression to achieve a seal. The comments on care during installation, under (1) apply equally to foam strip sealants.
3. Pre-formed polyisobutylene sealant strip with flexible liner and paper release backing. This is used to seal the joints on profiled metal liner sheets where these are intended to form the vapour control layer.
4. Low temperature sealant laminated onto a plastic film to provide a vapour barrier tape which allows movement. Aluminium foil backed tapes are available for similar applications.
5. Gun-applied compounds of the correct consistency for application to profiled metal sheeting. These gun-applied compounds are dependent on operator skill for the production of an unbroken bead of adequate size to seal the joints and all corners fully.
6. Grease sealants may be considered where profiled metal (normally aluminium) sheets nest closely at end laps creating a thin joint. With the wide range of available greases and additives used in their formulation, early contact with the manufacturer's technical department is essential to making the correct choice. Grease can leach out of end laps, attract dirt and look unsightly.

   Some greases contain metallic substances including graphite and zinc and their potential for contributing to bimetallic corrosion may be judged from their relative positions in the 'electromotive series', graphite being cathodic (at the 'noble' end) and zinc anodic (towards the base metal end).

One of the most widely used industrial greases is lithium-based and water resistant grades incorporating oxidation and corrosion inhibitors which are occasionally used with aluminium roofing. Such greases having a 'drop point' of 180°C and above, are considered suitable for use at surface temperatures likely to be encountered on a roof. They are available in cartridge form.

8.4 APPLICATION OF SEALANTS

Sealant should normally be applied first to the underlapping sheet, ensuring that it is continuous, of correct breadth and thickness and taken into all profile corners without 'bridging' (see Figures 8.1 and 8.2). When placing the overlapping sheet care must be taken to avoid disturbing the sealant. Compression of the sealant is essential to achieve a satisfactory weather seal.

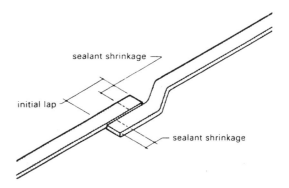

Figure 8.1 Sealant joint.

Ideally, sealant should be set back from the edge of an overlapping sheet just enough to give protection from the sun's UV rays, but close enough to the edge to minimize the amount of water held in the joint by capillary action.

8.5 PROFILE CLOSURES

8.5.1 Profile fillers

Also referred to as profile foam fillers, these accessories provide a means of closing the inter-profile cavities at ridge and eaves, where they are inserted between the roofing and ridge piece or eaves flashing (see Figure 8.3). They serve to exclude rain, snow, wind driven debris and birds, although the latter are known to damage the closures by pecking. Attack by birds can be minimized by using profiled metal trims to cover the foam fillers.

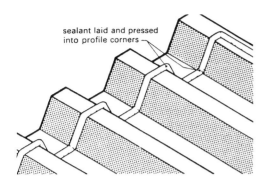

Figure 8.2 Sealant – continuous bridging at corners.

Available in a range of plastics material suitable for the relevant operating temperatures, they are capable of being compressed for retention and are also available with self-adhesive longitudinal edges.

Durability is claimed to be from 15 years for products of closed cell cross-linked polyethylene, to 20 years for closed cell EPDM (or polyethylene faced with EPDM), having UV resistance ranging from 'good' for the former material to 'excellent' for the EPDM products.

Water absorption during total immersion is stated to be 1% by volume for solid EPDM, rising to 4% for the polyethylene (marginally less when faced with EPDM).

8.5.2 Vented fillers

If passive ventilation of the sheeting profiles is necessary for the removal of moisture, closures with mesh inserts or of the cut-back or notched types must be used (or closures omitted and other measures used to exclude birds).

When closures are fitted between ridge capping and sheeting the ends should be butt-jointed over profile crowns, to minimize the entry of wind-blown rain if shrinkage causes open joints, for the same reason, ridge closures of the ventilated type should be fitted with their openings uppermost.

It is advisable for ridge closures to be set back by, say, 80 to 100 mm to discourage attention from birds (gulls are said to be particularly destructive). It is necessary, however, that the roof sheet continues for at least 100 mm towards the apex, behind the ridge closure to trap any minor leaks past the closure.

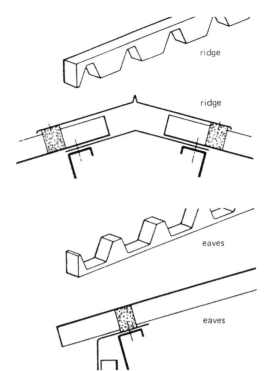

Figure 8.3 Profile closures.

8.6 EAVES CLOSURE FOAM FILLERS

A different attitude is needed towards eaves closures. When a breather membrane is fitted it is laid over the insulant and lapped to convey any condensate from the roof sheeting to the gutter. If the breather membrane is to fulfill its intended function, closures must be fitted in such a way that moisture can drain to the gutter unimpeded.

8.7 FIRE RESISTANT CLOSURES

Fire resistant closures are available for use where a cavity barrier is required, but their use would complicate ventilation and draining of the profiles. The current Building Regulations do not require cavity barriers where Class O insulation is fitted in contact with metal outer sheeting.

9

RAINWATER GOODS

9.1 INTRODUCTION

Roof drainage systems have a very important and fundamental role in the drainage and watertightness of a building. It is particularly important to correctly design and construct parapet, boundary wall, and valley gutters which, if they leak, will tend to do so into the building, which is unacceptable.

Eaves gutters are normally positioned outside of the building line but nonetheless correct design, construction and positioning can reduce the occasional overflow caused by wind driven rain. It is normally impracticable to design a roof drainage system to cope with extremely heavy rainfall events however infrequently they may occur.

A balance needs to be achieved in the design of the roof drainage system between the cost of the rainwater goods for an infrequent storm and the consequences of flooding into the building.

It is normally considered that an adequate design is one which has the capacity to deal with relatively intense storms that normally occur during thunderstorm conditions in the summer months.

Generally the sizing of gutters is dependent on three factors:

1. the area of roof to be drained;
2. the position, spacing and number of rainwater pipes;
3. space into which the gutter is to be fixed in boundary wall and valley gutter applications.

9.2 DESIGN CONSIDERATIONS

9.2.1 BS 6367: 1983 Code of Practice for drainage for roofs and paved areas

Design methods for the drainage of roofs are detailed in BS 6367: 1983 Code of Practice for drainage of roofs and paved areas. The principal elements are detailed below. BS 6367 is being revised by technical committee, CEN/TC 165 with prEN 12056–3.

As a guide for initial design checks on flow rates of gutters, outlets and downpipes the minimum rainfall rates should be taken as:

75 mm/h where the gutter will overflow outside the building;
150 mm/h where any overflow will be inside the building.

9.2.2. Design rates of rainfall

9.2.2.1 The frequency and severity of the short term intense rainfall is dependent upon a number of factors and is detailed in Appendix A of BS 6367: 1983.

Although the upland areas of northern and western Britain have a higher annual rainfall than the lowland areas it is the lowland areas that experience more frequent intense short duration rainfalls.

9.2.2.2 The degree of risk of the consequences of flooding into the building has to be decided and this, coupled with the period of protection in years required for the contents, is used in Appendix A of BS 6367: 1983 to determine the design rate of rainfall. For example, if the building is in the south of England and contains electrical equipment which is susceptible to water ingress and has a design life of, say, 30 years then from Appendix A design rate of rainfall 150 mm/h for a duration of 2 minutes may be used.

9.2.3 Wind

9.2.3.1 As rain is normally wind driven in Britain this can have an effect on the area to be drained.

9.2.3.2 Allowance for the effect of the wind is not required when designing flat roofs, but an allowance must be made where pitched roofs or vertical surfaces occur that are freely exposed to the wind.

9.2.4 Snowfall

9.2.4.1 It is generally not necessary to consider the amount of run-off from melted snow in sizing gutters but cognisance must be made of the ability of the gutter adequately to resist snow loadings.

9.2.4.2 Rainwater goods, particularly gutters, and outlets, may become blocked by frozen snow which can have detrimental effects during periods of thawing with windy conditions. This problem can be overcome by the use of snowboards to bridge across openings during heavy snow falls.

9.2.4.3 It may also be necessary to use snowguards to prevent the possibility of damage to persons or structures from the effects of sliding snow at the open eaves of a pitch roof.

9.2.5 Run-off

9.2.5.1 In that wind affects the angle of descent of rain, the area of run-off is affected by the pitch of the roof and any adjacent vertical surfaces. The method of determining the run-off and the rate of run-off is detailed in Section 3 of BS 6367: 1983.

9.2.5.2 It is normally assumed that rain falls at an angle of one unit horizontally to two units vertically and the effect on a pitched roof is to increase the drained area from the flat plan area to a greater area which is a function of the flat plan area and a fraction of the rise of

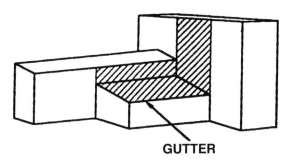
GUTTER

Figure 9.1 Shaded effective catchment area, A, is greater than roof area.

the roof (see Figure 9.1). For pitched roofs in the range 4–10°, the flat area can normally be used in the calculation.

9.2.5.3 If there are adjacent walls higher than the roof slope/roof drainage system under consideration then allowance must be made for any wind driven rain that will cause run-off from the wall into the gutter.

9.2.5.4 The required flow rate, Q l/s, according to BS 6367 for an effective catchment area, A m², is given by

$$Q = \frac{A \times I}{3600}$$

where I = design rainfall intensity in mm/h.

Where the flow of water into a gutter is not uniform along its length, the flows should be calculated for lengths of gutter between outlets with their appropriate catchment area.

9.2.6 General principles of design

The designed rainwater run off from the roof is conveyed into the underground drainage system by the following components:

1. the gutter which collects the run-off from the roof;
2. the outlets into which the flow in the gutter discharges;
3. the pipework that conveys the flow from the outlet into the underground drainage system (see Figure 9.2).

The general method of design is for the outlets and pipework to be of sufficient capacity to ensure that the gutter flows freely. Occasionally it is necessary to use outlets or pipework that is smaller than that required to ensure that the gutter flows freely and for this another, more complicated, design method has to be used. Both of these methods are described in Section 3 item 8 (Hydraulic design of roof drainage) in BS 6367: 1983.

Figure 9.2 Depth of flow in gutter.

The following assumptions are made for the general method of design.

1. The gutter slope is nominally level, i.e. it is not steeper than 1 in 350. If the gutter is laid to a greater fall then this can be considered as giving the design an additional factor of safety.
2. The gutter has a uniform shape in cross-section.
3. The outlets are of sufficient size to allow the gutter to discharge freely.
4. The distance between a stop end and an outlet is less than 50 times the upstream water depth or the distance between two outlets is less than 100 times the water depth.

9.2.7 Calculation of flows in gutters

The position of gutter outlets determines the flow along a gutter (see Figure 9.3) and should be used to determine the amount of flow.

The correct position of gutter outlets can have a considerable affect on the amount of flow in a gutter, e.g.

1. If the outlets, in a length of gutter, are at the ends then the flow along the gutter to each outlet is assessed to be half of the total flow.
2. If the outlets are positioned at quarter points, along the length of a gutter then, although the total flow in the outlet is the same, the flow in the length of gutter between outlets is halved.

For eaves gutters an outlet should be located near to each angle or change in the direction of flow (if this is possible).

For valley and parapet wall gutters the positions of an outlet must be such that the direction of flow is not changed sharply (e.g. through 90°) just before reaching it. A minimum of two outlets should be used where it is not possible to tolerate the gutter overflowing and it may also be necessary to provide for a weir overflow such that any flow in excess of the design rate is discharged to the outside of the building.

It is important in valley, parapet and boundary wall gutters to ensure that the overall depth of the gutter be greater than that required by the design capacity. This is

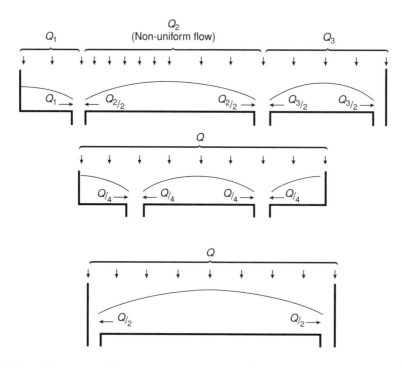

Figure 9.3 Unequal division of flow rate, Q l/s, between gutter outlets. Note that for the same flow, the gutter in (c) requires twice the capacity of the gutter in (b) (reproduced with permission of BSI).

to prevent the gutter being overtopped by splashing or by waves produced by strong winds. This excess of depth is known as freeboard and the following should apply.

1. Valley, parapet and boundary wall gutters – the amount of freeboard should be taken as two fifths of the maximum depth of flow in the gutter but not greater than 75 mm.
2. Eaves gutters – no allowance for freeboard need be provided if the overflow from the gutter falls clear of the building. A design flow rate of 75 mm/h is normally used for this condition.

The calculations of flows in gutters and outlets is detailed in item 8 of Section 3 of BS 6367: 1983.

Gutters may be of a standard type and size and the manufacturers will have tables detailing their capacities. Other gutters may be unique for the type of roof being used and will require to be designed in accordance with the above and will need to be manufactured in accordance with the requirements detailed in section 9.7.

9.2.8 Design of outlets

9.2.8.1 The gutter size and outlets need to be designed to accommodate the design rate of rainfall. The amount of rainwater entering the outlet dictates the type, size and positioning of the outlet. As a general rule at least two outlets should be included in any length of gutter, except for small external half round gutters.

9.2.8.2 Care must be taken in ensuring that the positioning of outlets within a gutter length is such that it is practicable to install the outlet at that designed position avoiding, for example, structural members. If the outlet has to be moved along the length of a gutter, from its designed position then this will alter the amount of flow into that outlet which could alter its ability to drain the gutter adequately.

9.2.8.3 BS 6367 recommends the use of box-receivers instead of rainwater pipe connections directly into the sole of the gutter, because they ensure free flow from the gutter and lower the risk of overtopping resulting from partial blockage (see Figure 9.4). The design rules provided for outlets of both types should be followed.

9.2.8.4 Where possible, tapered outlets should be used as these considerably increase the rate of flow into the rainwater pipes.

9.2.8.5 Side outlets should be avoided if possible as they are inefficient and consideration should be given at the design stage of the project to alter the design of, say, the steelwork to accommodate a sole box type

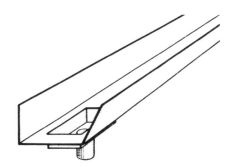

Figure 9.4 Box type receiver.

outlet. It is often the case that the position of a steel tie beam between columns precludes the use of a sole outlet, but it may be possible to alter this by consultation with the designer of the steelwork. If they cannot be avoided then box receivers incorporating the side outlet should be used.

9.2.8.6 The installation of weir overflows at both ends of a gutter is strongly recommended as they will give some warning of blocked outlets. Alternatively 'tell tale' pipe overflows can be fitted.

9.2.8.7 It is a common misconception that weirs can prevent gutters flooding. On long gutters weirs will not prevent flooding in severe storms, because the water flow along the gutter length is restricted by its width and flow characteristics.

9.2.8.8 Although gratings or wire balloons protect the rainwater pipe, they are a potential source of blockage and transfer the risk to the gutter itself. Gratings should be avoided unless the normal restriction to flow is taken into account in outlet sizing or the outlet exceeds 150 mm bore. Neither grating nor balloon is 'safe' unless gutters are regularly cleaned out.

9.2.8.9 Outlets from parapet wall and valley gutters must not be located near the downstream side of any point at which the water flow changes direction sharply.

9.2.9 Syphonic outlet systems

A number of syphonic systems of rainwater drainage are available, but the system design procedure is complex and will normally be carried out, supplied and fixed by specialist contractors. These systems will only have a syphonic action when the head of water at every outlet is sufficient to provide about 60% of that outlet's design capacity. At lower rainfall rates, and where one or more outlet has less head of water, the system will suck in air and work only as a normal gravity system, at significantly lower capacity. The depth of gutter should therefore be designed to accommodate storage of water between the syphonic pulses.

Syphonic drainage systems require that all the pipework from gutter down to the discharge point are airtight so that the syphonic suction can work. The discharge point is usually inside a manhole and it is essential that the maximum flow rate of water can freely discharge with no back pressure. Failures of syphonic systems have occurred where drain pipes under concrete floors have been damaged or the sewer capacity was too small. These parts of the system are normally outside the scope and control of the roofing or syphonic gutter contractor.

9.2.10 Gutter fixing

9.2.10.1   The use of correctly designed gutter straps is strongly recommended to provide adequate support for the gutter, unless the gutter is designed to be self-supporting. They should be placed at centres recommended by the manufacturer, but in any event one strap should be placed close to the socket at each gutter joint. It is recognized that gutter straps will be supported by the adjoining steelwork, and minor deflection, within tolerance, will occur. However if the deflection is excessive, i.e. more than 20 mm, then the steelwork contractor should be required to adjust the steelwork.

9.2.10.2   Straps used for valley gutters will interfere with the alignment of liner sheets and composite panels. Gutters which wrap over the purlins will not be adjustable, however they will provide a uniform line under the liner sheets.

9.2.10.3 Rainwater pipes of material requiring painting need to be spaced at least 30 mm from the wall face and this should be considered when positioning gutter or receiver outlets.

9.2.10.4   On built-up liner panel type roof systems, it is recommended that the liners, as well as the weatherskin, are detailed so as to drain into the gutter.

9.3 VALLEY, PARAPET AND BOUNDARY WALL GUTTERS

9.3.1   Gutter shapes

9.3.1.1   Where possible, it is advisable that cross-section shapes be used which comply with British Standards although, in many cases, they will have to conform to the cross-sectional shape of the roof and adequate depth is to be allowed so as to provide the required freeboard (see Figure 9.5).

9.3.1.2   Gutters should be large enough to enable pedestrian access for maintenance purposes and the recommended minimum width for a valley gutter is 500 mm and, for a parapet or boundary wall gutter, it is 300 mm. The design live loadings, to accommodate maintenance, need to be determined in order for the gutter to be designed to accommodate these loads adequately without undue deformation.

The gutter must also be capable of transferring its own self-weight and imposed loads due to rain, wind and snow to the supporting structure without undue deformation.

A method of achieving structural integrity is to use thicker gauge materials (e.g. 2 mm minimum galvanized steel and 3 mm minimum aluminium) with sufficiently sized gutter brackets at not greater than 1 m centres fixed to the supporting structure.

9.3.1.3   On particularly wide gutters in exposed conditions wave action has been experienced. To counter this, baffle bars can be incorporated spanning between the side walls. BS 6367 recommends a maximum freeboard of 75 mm, a minimum of 50 mm should also be considered to limit the risk of flooding from waves.

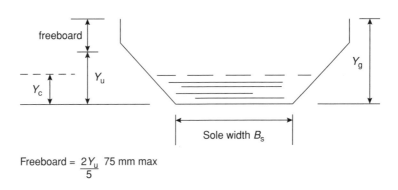

$$\text{Freeboard} = \frac{2Y_u}{5} \quad 75 \text{ mm max}$$

Figure 9.5 Typical valley gutter section: Y_g = gutter depth; Y_u = maximum depth of flow; Y_c = critical depth.

9.3.1.4 The design of the gutter is based on the gutter being level. Deflection of the gutter causing ponding should be avoided, if possible, as the ponding can reduce the flow capacity of the gutter and, in icy conditions, can make the gutter dangerous to walk through.

9.3.1.5 Water retention is not considered harmful to gutters submitted to hot dip galvanizing after manufacture. Ponding is unavoidable in many modern constructions where irregular steelwork is coupled with thin gauge materials.

Most gutters are supported, either by brackets or directly, by the structure (generally eaves beams or eaves purlins) and if these undulate or twist then the gutters will follow the same undulations.

This problem can be accommodated, to an extent, if adjustable gutter brackets are used but a hook-over gutter cannot take up any of the tolerance within the structure and is best avoided.

Gutter brackets or straps are recommended to provide a convenient means for adjustment where the straps will not interfere with the liner sheets. Straps have the advantage that they allow the gutter to expand and contract freely.

9.4 EXTERNAL GUTTERS

9.4.1 Eaves gutters

9.4.1.1 As water discharges from a roof edge it does so in a spread pattern, dependent upon the shape of the edge, the rate of run-off, the type of roof covering used and the pitch of the roof. The gutter should be positioned centrally under the roof edge and close beneath it.

9.4.1.2 Available in various cross-sections (half round, box or ogee), the gutters should be fully supported on brackets at 1 m centres (max.) with added support at outlets and changes of direction. Gutters also need to be restrained against wind uplift. A bracket should always be located at the socket end of a length of gutter to reduce loading on the joint .

9.4.1.3 Lightweight gutters may be supported from the roof or vertical sheet metal cladding using reinforcing plates or corrugated supports fixed to the roof sheets in accordance with the profile manufacturer's recommendations.

9.4.1.4 Provision must be made for the relatively high rate of expansion of aluminium, GRP and PVC. The outlets will form fixed points. Where an outlet at each end of a run is possible, consider extending the gutter in

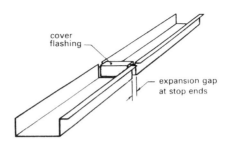

Figure 9.6 Gutter joints with stop ends.

both directions from stop ends at the mid-point, separated by an expansion gap and bridged by a saddle flashing (see Figure 9.6).

9.4.1.5 On buildings where design dictates that gutters run continuously around all four sides with purpose made corner pieces, it is recommended that expansion joints are incorporated along each side. The number of expansion joints and the distance between them is dependent on the size of the building and on the type of material used and therefore the manufacturer's advice should be sought.

9.5 GUTTER JOINTS

9.5.1 Methods of joining

9.5.1.1 Joints take the general form of a spigot and socket varying in complexity according to basic gutter design and may incorporate preformed sealing strips or rely upon gutter sealant. On steel gutters the joint is normally joggle type up to 3 mm thickness and typically a welded strap on thicker. Loose butt straps are not normally recommended.

9.5.1.2 Bolts used are normally M10 × 25 mm long set at average 75 mm pitches, 2 × 1.8 mm thick washers and one No. 3.0 mm thick Neoprene washer. Note that the Neoprene washer is located on the 'water side' of the assembly. Mastic sealing compound is usually 50 mm wide × 6 mm thick and based on a rubberized mastic to resist chemical attack.

9.5.1.3 The number of fasteners will be determined by the gutter manufacturer according to girth, shape and gutter materials and as a general rule washers should be fitted inside and outside. Bolting together of gutter joints should start at the centre of the sole plate and proceed outwards.

9.5.1.4 Fasteners for aluminium guttering should be of austenitic stainless steel. Aluminium gutters should be degreased, surface roughened and primed on the jointing surfaces.

9.6 GUTTER INSULATION

9.6.1 Thermal requirements

9.6.1.1 All internal gutters must be thermally insulated following the same principles as explained in Chapter 1. This means taking the same care with vapour sealing and the avoidance of thermal bridges. The integrity of the vapour control layer and air leakage barriers must be maintained throughout the roof and gutter areas.

9.6.1.2 It is recommended that gutters have less insulation than the roof so that snow and ice melt preferentially in the gutter. This can usually be achieved without lowering the overall insulation value of the roof envelope.

As a guide use 0.6 U-value for insulation on gutters with 0.45 U-value roof.

9.7 MATERIALS FOR GUTTERS AND RAIN-WATER PIPES

9.7.1 General

9.7.1.1 It is important to ascertain the design life required for the roof as this can influence the choice of material, sealants and fixings.

9.7.1.2 The cost and probability of maintenance being carried out also need to be assessed, as these can influence the choice of gutter material.

9.7.1.3 Cost cutting exercises at design stage are not always compatible with the size and shape of gutter eventually ordered.

9.7.2 Steel

9.7.2.1 All light gauge metal has a natural wave, undulation or bow in the sheet even before manufacturers process it. This nominal deformation can be further accentuated by bending, welding and hot dip galvanizing. The degree of waviness may also be enhanced by the chemical composition of the steel and the degree to which its thickness varies throughout the sheet. Such nominal distortion is inherent in the sheet, which is why the right gauge or thickness of material relative to the gutter size is crucial in design. Gauges selected which are too thin for the gutter size are processed at customer risk and any resulting distortion would not be considered

grounds for rejection. From experience, gutters with soles exceeding 400–500 mm in 2 mm may start to deform. Further, note that material thicknesses quoted are nominal, i.e. 2 mm may vary by ± 0.24 mm.

EN's are being written on gutters and will shortly replace the relevant British Standards. These are EN 1462, EN 607 and others.

9.7.2.2 Where possible it is recommended that galvanized steel used for gutters should be in accordance with BS 1091: 1963, hot dip galvanized after manufacture (GAM) to BS 729, nominal 3 mm thickness, to give full weather protection.

9.7.2.3 Galvanized steel which is hot dip coated with 600 g/m² of zinc in long coiled strips before manufacture (GBM) is widely used. GBM material is susceptible to immediate rust problems where cut edges, punched holes, and welding expose bare metal. The thickness of zinc varies depending on which grade of steel is used, the following is typical:

> BS EN 10241 DX51D + Z600 spangle finish, minimized spangle, bright matt, matt.

9.7.2.4 Dimension tolerances
The following are applicable to GBM, GAM and precoated light gauge steel.

Up to 6 m long gutter sections	–0.00 mm + 6.00 mm
Up to 1.5 m girth	–0.00 mm + 3.00 mm
Bends (dimension between)	± 1.00 mm
Bolt hole centres	± 1.00 mm
Angles	± 1°

9.7.3 Aluminium

9.7.3.1 Aluminium grade (99% pure) used for bending should be in accordance with BS EN 494. Sometimes aluminium gutters will be polyester powder coated. It should be noted that any surface welding carried out will show through on the reverse side of the sheet as rippled heat marks. Such marks will also show through paint.

9.7.3.2 To avoid bi-metallic corrosion only austenitic stainless steel fixings should be used. Further note that in certain (high saline) conditions aluminium will rapidly deteriorate by turning to white powder, or 'chalking'. Aluminium should be protected from the effects of corrosion by contact with or run-off from copper, cast iron and steel (including grades of stainless steel – notably the low chromium, non-austenitic types which are less able to form a strong self-protecting surface film), alkaline cement materials and plaster.

9.7.4 Coatings for mild steel and aluminium

9.7.4.1 Polyester powder coatings were devised for coating aluminium, they can now be applied to galvanized finish. The bumps and runs in zinc coating show through paint finish and are not always aesthetically acceptable. Polyester powder coatings are often specified by a trade name although similar product performance is now available from other suppliers.

9.7.4.2 Black bitumen-based coatings or similar are required on all steel gutters. These should be applied after about 12 months weathering of the new galvanized steel gutter and need regular re-applications to remain effective.

9.7.5 Stainless

Austenitic grade should be used, the ordinary commercial quality is dull grey; the catering quality has a mirror finish. Despite the name, it marks easily and when contaminated with mild steel dust will show rust staining very quickly. The cost is approximately four to five times that of mild steel.

9.7.6 Plastisol coated steel

This is generally manufactured from 0.7 mm thick steel with plastisol coatings 200 μm thick to both sides. This lightweight type of gutter is generally supported from the roof sheets by support arms, sealed and fixed over the corrugation of the roof profile. However, systems can vary, and therefore consult the manufacturers. Also it is only considered suitable for external eaves type application – not for valley/parapet gutters, i.e. non-trafficable areas. Light gauge gutters are typically jointed using a loose butt strap, either internal or external. Fixings can be either M5 stainless steel nut, bolt, washer and bonded Neoprene washer complete with colour cap, or sealed rivets complete with caps. A seal is achieved with a soft butyl mastic or silicone sealant.

9.7.7 Glass fibre reinforced plastic

9.7.7.1 Glass fibre reinforced plastic (GRP) components usually use polyester resin giving a structurally durable product with a strength to weight ratio which is superior to that of mild steel and is capable of being pigmented to give internal colour. Colour fading is likely within a few years of exposure to the elements, but will vary with pigments. Weathering of the surface is improved by the use of a resin gel coat which reduces the risk of the fibres being exposed.

9.7.7.2 GRP is suited to double-skin construction, allowing thermal insulation to be completely enclosed within two layers. Attention should be paid to the manufacturer's recommendations for expansion joints.

9.7.8 Composite gutters

If gutter systems are being used which require the use of different materials, e.g. elastomeric sheets prebonded to a metal gutter but with site formed jointing, then adequate training of the operatives is strongly recommended. Gutter linings to the weather side may be formed from malleable metals such as lead, aluminium, zinc and copper, all of which must be considered for potential bimetallic corrosion. Other materials for built-up linings include bitumen-based products and synthetic elastomers in sheet form, some of which are virtually unaffected by the effects of sunlight, oxygen and ozone, but these await long term evaluation.

9.7.9 Unplasticized polyvinyl chloride

Unplasticized polyvinyl chloride (uPVC) has a high coefficient of linear expansion for which due allowance should be made. Also, because of low temperature embrittlement, it requires careful handling in winter conditions. This type of gutter is generally supported by purpose made brackets fixed to fascia boards. (See BS 4576: Part 2, Half-round gutters and circular pipe.)

9.7.10 Fibre-cement

Fibre-cement rainwater products have now generally been substituted as replacement for asbestos cement products. No British or European Standard is currently available for these products.

9.7.11 Asbestos cement

Asbestos cement is *not generally available*. However, its use is subject to dust-control and safety procedures covered by the provision of the current Asbestos Regulations.

9.7.12 Cast iron

The inside surface of cast iron gutters needs painting for protection. Run-off from aluminium roofing can cause corrosion. (See BS 460: 1964, Specification for cast iron.)

9.7.13 Outlet protection

If wire balloons are used to protect outlets from blockage, they need to have adequate design life and be of such a size that they cannot be pushed down into the outlet. The wire balloons are likely to need replacement within the design life of the gutter.

9.8 INSTALLATION

9.8.1 Insulated gutters

When using insulated gutters it is particularly important to ensure that site formed insulation is carried out with due care being taken to ensure that cold bridging is eliminated and that adequate support of the insulation is made. Care must also be taken at joint positions to ensure that there are no gaps in the insulation.

9.8.2 Training

If gutter systems are being used which require the use of different materials, e.g. elastomeric membranes pre-bonded to a metal gutter with site formed joints, then adequate training of the operatives is essential.

9.8.3 Final checking

A complete check on all component parts of the gutter installation should be made prior to leaving site but in particular the following must be checked:

1. that all fixings have been installed and are adequately tightened;
2. that all patch outlets that are site installed have had all burrs removed from the edge of the gutter in order not to provide a place on which debris can lodge and also that any damage to the parent metal caused by cutting the opening in the gutter has been correctly treated to prevent any possible corrosion;
3. that all debris has been removed that could cause a potential blockage of the gutter;
4. that any damage to a gutter surface is repaired to prevent a weak spot being formed at a later date at which corrosion can occur.

9.8.4 Workmanship

9.8.4.1   In general, corrosion resistant gutter bolts, nuts and washers should be used, tightening first the fixings in the sole and working towards the sides with the nuts accessible after assembly for possible future adjustment.

9.8.4.2   Where safety considerations allow, it is recommended that the direction of lay should be away from the outlets. After laying the mastic in the joggle or butt strap, ensuring that it sits in the corners and is not

'stretched across' them, tighten the bolts in the sole of the gutter first, and then work evenly up both sides. The bolts should not be tightened as far as is possible, but only sufficient to compress the mastic by 66%, i.e. 6 mm mastic reduced to 2 mm, the excess being removed.

9.8.4.3   If jointing aluminium gutters using preformed seals, follow the manufacturer's instructions and, if using an approved mastic, the joint surfaces should be prepared by removing any grease or oil film, roughening and applying a metal primer compatible with both mastic and the aluminium.

9.8.4.4   During the execution of roofing work, gutters must be cleared of all debris at regular intervals and due care and attention given to the avoidance of damaging or defacing the gutter surface or its fixings.

9.9 STAINING

Rust staining is a common occurrence in steel gutters because, being the lowest point on a roof, all the roof debris will eventually wash into them. Metal dust and swarf from drilling are the commonest cause. It is therefore good practice to sweep gutters clean before leaving site. Surface staining is not harmful provided the material causing staining is removed and not left to lie.

9.10 LEAKS

The commonest causes of leaks are:

- contaminated joint compound;
- fixing during wet weather;
- mastic pushed out of joggle during fixing;
- split or damaged washers;
- bolting sequence erratic, i.e. not from centre of sole;
- cross-threaded bolts, bolts not tightened;
- bitumen paint melting the joint sealing compound;
- heavy foot traffic in light gauge unsupported gutters;
- gutters used as (over-loaded) storage platforms.

9.11 MAINTENANCE

It is recommended that clients are advised in writing that all gutters should be inspected at least four times a year, cleaned out, outlets checked and re-coated as necessary. The designer and roofing contractor have obligations under CDM regulations to consider how the gutters can safely be maintained. By experience it may be possible to reduce the inspection frequency to twice per year where there is minimal risk of blockage caused by leaves, etc., see also Roof Maintenance in Chapter 10.

10

CONSTRUCTION, SAFETY AND MAINTENANCE

10.1 INTRODUCTION

The widespread acceptance of roofing/cladding systems which combine into one sub-contract the operations of weatherproofing, insulating and lining and feature pre-finished materials for external and internal decoration, has brought new opportunities but also new problems.

At no time can there have been greater pressure on management to ensure the full utilization of resources. On the financial side, the welcome increase in sub-contract value is frequently associated with the use of more costly materials, creating the risk of equally costly wastage. Of greater importance is the potential for expense to be incurred in the investigation of complaints of roof leakage and of remedial action if condensation is finally diagnosed and the roofing contractor is unfortunate enough to be responsible under the terms of the sub-contract!

The fact that technical staff face problems of increasing complexity is evident from the content of several relevant British Standards which have recently been revised or are currently undergoing revision. New products often demand fresh skills from the workforce and changing techniques are likely to involve new tools and equipment. New safety legislation has added to the responsibilities of roofing contractors.

These factors highlight the continuing need for a commitment by management to training and, whilst this calls for the provision of financial resources, failure to make a positive response will undoubtedly prove the more costly option in the long run.

Training is not for the new entrant only. 'Experience' is gained in the past: with re- training it looks to the future.

10.2 SITE SAFETY

Remember ignorance of the law excuses nobody!

10.2.1 The Health and Safety at Work, etc., Act 1974

The Health and Safety at Work, etc., Act 1974 imposes statutory duties both on employers and employees.

The aims of the Act as set out in Section 1 are:

1. securing the health, safety and welfare of persons at work;
2. protecting people who are not employees at a particular place of work against risks to their health or safety arising out of, or concerning, the activities of the employed people working at that particular place;
3. controlling and keeping the use of explosive, or highly flammable, or otherwise dangerous substances and generally preventing the unlawful acquisition, possession and use of such substances;
4. controlling the emission into the atmosphere of noxious or offensive substances from premises belonging to all people covered under the Act.

10.2.2   Employer duties

The Act imposes certain duties upon the employer:

- to provide and maintain plant and systems of work which are as far as practicable, safe and without risk of health;
- to make arrangements for the safe use, handling, storage and transport of articles and substances;
- to provide a safe place of work including safe means of access and egress;
- to provide a working environment that is as far as practicable, safe and consistent with the welfare of the employees.

These duties are supplemented by a general duty to non-employees (e.g. visitors) at a place of work.

10.2.3   Employees' duties

The Act also imposes duties upon employees who, whilst at work must:

- take reasonable care for personal health and safety of themselves and other persons who may be affected by any acts or omissions;
- co-operate with the employer in matters of health and safety;
- not intentionally or recklessly interfere with anything provided in the interest, safety or welfare of employees.

10.2.4   Safe systems of work

Safe systems of work can only be achieved and maintained through careful site planning and the adoption of correct working procedures. Supervisors and operatives must be made aware of their responsibility for employing safe methods of working and adapting to changing circumstances when moving from site to site.

Safety training is an essential part of staff development and it is essential for operatives to receive approved instruction in the safe use of abrasive wheels, cartridge operated fixing guns, power tools in general, lifting gear, access equipment, etc., and in the precautions to be exercised when working with the wide range of materials and substances encountered in the course of their work.

10.2.5   Existing roofs

Existing roofs should always be treated with extreme caution and roof boards used. Prior to stripping, a survey should be carried out to evaluate any risks from hazardous processes within the building, any deterioration of the roof through corrosion, or harmful substances which might be disturbed during the work. Where asbestos is found or suspected then the current regulations concerning working with asbestos fibres must be implemented.

Great care should be taken to establish if existing rooflights are safe or if any have been concealed by over-coating with opaque substances.

Any such disclosures must be taken into account when defining safe working procedures for the contract.

10.2.6   Access

All working areas must have proper means of access, proper landing platforms and ladders of the correct height as laid down in the Construction Working Regulations, which should be adhered to at all times. Edge protection and handrails are mandatory, safety harness should be made available and used wherever necessary.

Passive safety methods are preferred to those where active participation is necessary such as when using safety harnesses. Safety nets rigged below areas where roof sheeting is being fixed to purlins are currently recommended by HSE in HSG 33: 1998 as a suitable form of passive safety measure.

Access around the perimeter of the building is important for the movement of materials and scaffolding, especially when roofing and cladding is being undertaken.

Consideration should be given when working on roofs, to the safety of those who might attempt to pass beneath. An area should be cordoned off and adequate means of warning or protection given. Protection and working platforms should never be used as load bearing (e.g. storage) areas unless specifically designed as such. Roof areas and walkways should always be kept clear of debris and stacks of material to be fixed, or having been removed, must be fastened and made safe.

10.2.7   Plant and scaffold registers

Wherever plant and scaffolding are provided it is the law to keep detailed registers of inspection and to have a system of handover certificates for use when scaffolding is altered, moved or accepted. The maintenance of plant and any necessary procedures for its inspection, testing and certification must be carried out at the requisite intervals.

10.2.8 Site safety meetings

On a well organized site the main contractor will convene a safety meeting at the commencement of the contract and other similar events for sub-contractors as they are brought in. These should give an opportunity for safety matters to be aired and procedures discussed and are often a timely reminder to operatives of their duties. Operatives should be encouraged to attend and record sheets should be endorsed by the main contractor's representative to confirm that they did so (the signature is usually given 'for record only' but that is what is needed!).

If no such action is taken by the main contractor, safety procedures adopted by the sheeting sub-contractor should still be monitored and reviewed at regular meetings called by a supervisor or other responsible person.

10.2.9 Welfare

Welfare is concerned with providing at least the basic level of hygiene and comfort by way of mess room, drying room, first aid facilities, etc., and includes services which often do not receive the attention they deserve, e.g. induction of new operatives, assistance with the form-filling bureaucracy and an understanding of personal and domestic problems.

Improved facilities are not always taken for granted and can lead to better working relationships and attitudes towards safety to the benefit of the individual and company alike. If better facilities help to improve the unhappy record of the construction industry in respect of fatal accidents and serious incidents the effort will be worthwhile.

10.3 HEALTH AND SAFETY SECTION

10.3.1 Health and Safety Executive publications

For everyone connected with the design and application of profiled sheeting for roofing and cladding, it is essential for them to be familiar with and apply the guidance given in the Health and Safety Executive publications HSG33, *Health and Safety in Roofwork* and HSG151, *Protecting the Public*.

The following summary indicates some of the laws on health and safety which cover all work activities.

10.3.2 The Health and Safety at Work, etc., Act 1974

The Health and Safety at Work, etc., Act 1974 (HSW Act) imposes statutory duties on employers and employees. This Act applies to **all work activities**. It requires employers to ensure the health and safety of their employees, other people at work on site and members of the public who may be affected by their work. If they employ more than five people, employers should have a clear and simple written down Health and Safety Policy. Employees have to co-operate in the

implementation of this policy and be trained in understanding and applying the policy.

10.3.3 The Management of Health and Safety at Work Regulations 1992

The Management of Health and Safety at Work (MHSW) Regulations 1992 apply to everyone at work and require employers to plan, organize, monitor and review their projects by carrying out all necessary risk assessments, have competent health and safety advice available, and provide information and training to all employees to fulfill the requirements of the Act of 1974. The MHSW Regulations 1992 require risk assessments, for companies employing five or more persons, to record the significant findings of the assessment and any groups found to be specifically at risk.

10.3.4 The Construction (Health, Safety and Welfare) Regulations 1996

The Construction (Health, Safety and Welfare) Regulations (CHSW Regulations) 1996 apply to **all construction work** that is carried out by a person at work and upon all construction sites including domestic works. It includes special provisions for higher risk trades such as roofing.

For example, it states that when a working platform requires edge protection, that protection should be to the following specification: Toeboard – minimum height 150 mm; Main guard-rail – minimum height 910 mm; Intermediate guard-rail (or similarly secured alternative) to ensure no gap exists which is greater than 470 mm.

When the risk of falling is assessed, it may be necessary to put in place guard rails and toeboards even if the potential drop of the fall is only 1.5 m, if the fall were to be over starter bars or other high risk areas. If the potential drop is 2 m or more, the full level of edge protection or other similar means of protection is always required. Where working on the leading edge of the roof system, or on a moving working platform along a roof or any area where falls are not reasonably practicably protected by guard-rails, then personal protection, i.e. harness and line or in some cases nets, must be introduced unless the risk from such a fall is minimal.

Any person who will be working near, on, or pass across a fragile material must be provided with adequate protection to prevent a fall and there must be warning signs in the approach to the area stating the existence of fragile material.

10.3.5 The Construction (Design and Management) Regulations 1994

The Construction (Design and Management) Regulations (CDM) 1994 apply to construction work, including roofwork when it is expected that the work will take more than 30 working days or 500 person days, or if it is expected that there will be five or more workers on site at any one time or if any demolition is involved.

Where work is carried out for a domestic client, contractors must notify HSE if the work will take 30 days or more and designer's duties apply, such as taking account of buildability, maintenance and repair issues within their design.

All the above regulations, along with others, apply to all sub-contractors, including public utilities and the self-employed. Sub-contractors will be required to provide method statements and risk assessments for the area of roofwork to be carried out. This will require the roofing contractor to develop and be able to implement and show competence in method statements, using safe systems of work, and in preparing risk assessments which heed the company's written Health and Safety Policy prior to their appointment to carry out works. Thus, a large element for sub-contractors is the need to be able to provide proof of competence to undertake their work within the CDM parameters.

However, roofers and designers should note that a roofing contractor becomes a designer when circumstances such as the following occur.

- The roofing contractor is offering a design and build package; or
- when selecting and/or ordering the individual parts or a combination of profiled sheets, spacers, insulation, VCL products, breather membranes or linings; or
- when altering the specification (e.g. by providing another system equal to the product specified); or
- when the products and systems have not been chosen/specified by the client, his designer or the main contractor and the roofing contractor is asked to provide an adequate system.

All designers have a duty to inform their clients, excluding domestic, of their legal duties under the CDM Regulations. This roofing contractor must collaborate directly with the planning supervisor in the creation of the Health and Safety Plan for the project and in providing the necessary risk information on all products and systems used, for inclusion in the Health and Safety File (which is only concerned with any significant risks to be taken into account during the life of the building after the completion of the construction phase).

The arrangements necessary to comply with the extra duties and responsibilities which a roofing contractor accepts, by carrying out designer functions by choosing materials and/or systems, are outlined in the HSE CDM 1994 Construction information sheets Nos 41, 39, 43 and 44, *The role of the designer, The role of the client, The health and safety plan during the construction phase*, and *The health and safety file*, respectively. Formal enforcement of CDM to bring about compliance might include the issuing by HSE of Improvement Notices or Prohibition Notices.

We suggest that any method statements offered to the principal contractor or to the planning supervisor by the roofing contractor/designer should include reference to the guidance on site roofing operational conditions contained in the NFRC's 1997 revised edition of *Roofing and Cladding in Windy Conditions*.

The following regulations must be included in your health and safety policies, arrangements and method statements to enable you to demonstrate compliance with the law:

The Health and Safety (Consultation with Employees) Regulations 1996
The Reporting of Injuries, Deaths and Dangerous Occurrences Regulations (RIDDOR) 1995
The Workplace (Health, Safety and Welfare) Regulations 1992
The Construction (Head Protection) Regulations 1989
LOLER Lifting operations 1998
HSE construction information sheets Nos 17 (revised), 20 and 22 *viz. Construction health and safety checklist, Short duration work on pitched roofs* and *Work on fragile roofs: Protection against falls*
BS EN 1263-1: 1997, Safety nets Part 1: Safety requirements, test methods
BS EN 1263-2: 1995, Part 2: Safety requirements for erection of safety nets

It is anticipated that following the BS EN's 1263 Part 1 & Part 2 – Safety nets, a revised BS 8093 – "Code of practice for the use of safety nets, containment nets and sheets on construction works" will be published. Safety nets can be effectively employed to minimise the effect of falls in roofwork. For profiled sheet roofing, safety nets professionally designed, manufactured to BS EN 1263 and installed, offer collective, passive safety as they protect everyone working within their boundary. Type 'S' Safety nets, described in the BS EN as nets with border rope, are becoming known as the preferred option on sites using profiled sheet roofing, since they do not inhibit the range of activity of operatives within the net's boundary. If there is a fall, the high energy absorption capability of a safety net should ensure a soft landing but it depends on the net being fitted as closely as possible to the underside of the working platform. It should be possible to position the net so that at the point of maximum sag, it is less than 2m from the roof surface.

The manufacturer's recommendations should be followed on the number and spacing of fixing points and checks should be made to ensure that the supporting structure can resist the expected anchorage loads.

Safety net attachment points should be included in the primary steelwork so that safety nets can be used during erection and subsequent maintenance, alteration and eventual dismantling of the profiled sheet materials. Only operatives trained and competent in the erection of safety nets should carry out this work. Specialist installers and riggers are likely to be required on many projects and certainly where large nets are to be used.

10.4 TRANSPORT, HANDLING AND STORAGE

10.4.1 Transport

Knowing that arrangements vary from one contract to another and that such matters can not be left to chance, the issue of responsibility for off-loading and storage of materials should be agreed before delivery takes place.

The supplier has a duty to inform the sheeting contractor at an early date of any special techniques or equipment needed to handle a given consignment safely and these requirements must be passed on to the party responsible for putting them into effect on site.

Make sure you get the information but don't keep it to yourself!

Once profiled sheets have been erected as a roofing or cladding in the outside atmosphere, they have a life of many years. However, failure to observe simple but essential precautions when storing and handling cladding sheets can lead to damage, delay and expense. It can also affect the long term performance of a cladding system.

Bulk deliveries should be packed and presented in a manner suitable for off-loading by mechanical means, whilst access to and on site must be adequate for the operation of fork-lift trucks or mobile cranes as necessary for the unloading of heavy crates or pallets of metal sheets delivered on long trailers.

Suppliers should act to encourage the correct off-loading procedures.

The packaging must ensure protection of edges and corners against damage in transit and, with normal handling, guarding against abrasion due to movement and preventing surface staining by rain. New European product standards will require suppliers to comply with recommendations for labelling which any reputable manufacturer should already have adopted. The label on each pack should at least identify the manufacturer, order number and contents. Additional information such as intended (grid) location on the roof or wall may be required by the roofing contractor.

When it is necessary to control the deflection of long unit loads, suppliers should see that the approved lifting points are clearly marked on each consignment together with the gross weight, to encourage the use of suitable lifting gear.

10.4.2 Handling

Sheets in bulk should be handled using a lifting beam in conjunction with suitable slings and spreaders to avoid crushing or chafing the sheet edges, or by fork-lift truck with appropriate fork dimensions (see Figure 10.1).

Insulated panels should be handled strictly as directed by the supplier and, following all deliveries, site supervision should inspect the materials, record any shortages or damage in transit and without delay inform their contracts department.

10.4.3 Storage

Provision should be made for secure storage under cover of 'small' items, whilst a firm, level, dry base protected from the weather, accidental damage and theft is required for metal sheets. Supplier's instructions must be followed when protection from heat, sunlight or low temperature conditions is required.

Unless delivered on their own integral pallet, sheets should be stacked on bearers spaced at intervals of not more than 900 mm and set level or to a fall as directed by the supplier. Stacking heights will vary according to product, and again the supplier should give instruction (see Figure 10.2).

Storage space should be allocated as near to the point of usage as possible and stacking arranged to give access in the order of requirement for fixing.

Before sheets are placed upon the roof structure for storage in bulk the steelwork contractor must be consulted and advice followed. Any such agreed storage must be properly secured against all foreseeable contingencies.

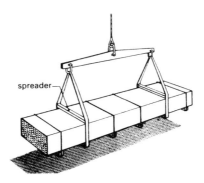

Figure 10.1 Lifting crane using spreader beams.

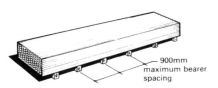

Figure 10.2 Storage bearer spacing.

If storage indoors is not possible, protect the sheets with a tarpaulin or waterproof covering, which should be supported on a scaffolding frame, leaving sufficient room on all sides for air to circulate. Store the sheets off the ground on timber bearers, which should be spaced no more than 900 mm apart. Incline the stacks, so that any rain which penetrates the covering will drain off. Failure to keep metal sheets dry in a pack can result in staining of aluminium and white rust on galvanized steel.

Advice on the storage of rooflights, especially factory insulated units which can be easily damaged before installation, are given in Chapter 6, Rooflights.

Inspect the sheets at regular intervals to check for leaks in the covering. Never drag sheets from a stack. Carefully lift and turn them. Carry them on edge. All materials should be handled with care. Insulated panels, especially if faced with a foil vapour barrier, are vulnerable and should be treated with special care.

Wherever possible lift single sheets manually on to the roof, if a sheet has to be hoisted into position, make sure that its edges are protected and that pressure across the sheet does not distort it. Use only ropes or slings for hoisting – never chains.

Never store sheets where they may be walked on. When working on a roof wear only soft soled shoes with a firm grip.

Care at this early stage will be rewarded by sheets that look good and perform well.

10.5   PREPARATION

10.5.1   Commencement of work on site

Before commencement of work on site the senior sheeter must be fully briefed, with reference to the use of all relevant drawings and details. The briefing should include reference to contractual arrangements for off-loading, scaffolding, hoisting, waste disposal, messing facilities, attendance at safety meetings, etc.

In conjunction with the supervisor, the senior sheeter should check that all materials, components and sundries for a particular area or phase are on site and in fit condition for use and that the plant requirements have been satisfied.

10.5.2   Additional site hints

Check sheet colour finishes for possible 'grain' effect, they may need correct orientation for a uniform appearance! Where possible avoid mixing sheets from different batches on a wall or roof area.

The importance of correct sheet alignment at the commencement, can not be over-emphasized (see Figure 10.3). Roof sheets must be set at right-angles to ridge/eaves lines and to an eaves datum line. Wall sheets must be plumb and set to a lower edge datum line. All laps are to be correctly engaged and any tolerances on sheet length should be accommodated under end laps and unseen.

Where recommended by the material manufacturer, use roof boards to protect roof sheeting from foot traffic. Temporary protection against foot traffic is recommended for GRP and membrane coated valley gutters.

Keep surfaces free from swarf and other debris.

10.6   MAINTENANCE

10.6.1   Maintenance manual

A maintenance manual should be provided to the building owner or responsible party on completion of the roofing and cladding work. This is often included with the Health and Safety Manual for the building.

The following items should be considered for inclusion in the maintenance manual:

1. any additional safety information not included in the Health and Safety Manual;
2. copies of all relevant product manufacturers' guarantees and recommendations for maintenance;
3. many product guarantees require regular inspection of the roof and wall cladding even during the incorrectly named 'maintenance-free period'; some

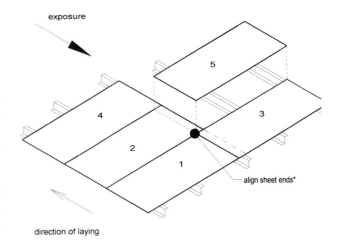

Figure 10.3 Sheet laying sequence.*Cover any variation in sheet length under lap (not possible with composite panels).

Table 10.1 Check-list for inspection of roofs (from BS 5427: Part 1: 1996)

Item	Action
1 Leaks, inspect from inside for location	Repair as soon as possible and do not leave until annual maintenance
2 Damage or decay	Repair, or replace, as necessary
3 Accumulated debris, e.g. trapped leaves or pine needles	Clean away
4 Gutters and drain pipes. To be clean and free draining	Clean out
5 Gutter joints. Inspect for defective sealant or loose bolts	Repair
6 Loose ridge or other flashings	Repair
7 Missing foam fillers or evidence of damage by birds	Replace
8 Staining caused by ponding or water	Seek advice from roofing specialist
9 Discoloured areas of the surface coating	Inspect the surface coating and attend to any peeling, corrosion or other evident deterioration, as necessary
10 Discoloured areas around fume extract ducts should be closely inspected	Re-coat without delay where attack of the coating has occurred
11 Evidence of access to the roof, e.g. for maintenance of ventilators	Touch-up any scratches in coating
12 Missing colour caps from fasteners	Replace
13 Missing fasteners	Investigate and replace
14 Corroded or degraded fasteners (including sealing element)	Seek advice on replacement *Note:* Fasteners should not normally need to be re-tightened but where it is necessary due to vibration or other causes. Special care should be taken not to compress the insulation below the fixing.
15 Sealants visible on roof surface	Seek advice on replacement and suitability of the sealant

require that an inspection record form is completed and returned for their files;

4. a sample record sheet for recording periodic inspections of the roof and wall cladding;

5. a copy of the recommendations and inspection checklist for roofs from BS 5427: 1996, which is reproduced in section 10.6.2;

6. advice on sources of suitable materials for maintenance which should include details of when to use, e.g. timing of the decision to repaint coated metal after the original guarantee has expired.

10.6.2 Maintenance inspections

The following advice is reproduced from BS 5427: Part 1: 1996:

'Inspection should be carried out by suitably experienced persons. All roofs should be visually inspected at least once per year and always after severe storms. Lack of maintenance will accelerate deterioration.

Where practicable, roofs should be inspected from the eaves or gable ends, avoiding the need to walk on the roof. When access to the roof surface is essential protective boards should be used unless it has been established from the sheet manufacturers that the sheet can be walked upon.

The checklist given in Table 10.1, although not exhaustive of all possibilities, is a guide to the minimum; similar actions apply to wall cladding where appropriate.'

10.7 FAULT FINDING ANALYSIS

The point has been made earlier that the 'likelihood and extent of condensation are largely determined when the specification is prepared'. It is perhaps a client's natural reaction to regard the appearance of moisture as a leak and point the accusing finger at the sheeting contractor.

Whilst no roofing contractor would wish to convey to the client the impression of wishing to 'pass the buck', it must be in the interests of self-preservation to be able to differentiate between a leak and the result of a possible design weakness.

Thus although analysing faults in profiled metal roofing is time consuming it must be carried out systematically, and remedial action commenced only when the cause and location of the problem is known and degree of responsibility established.

The following suggestions are put forward as a recommended course of action after condensation has been eliminated as a cause of any reported leaks.

1. Obtain plans of the building showing, if possible, the purlin layout and any special roof features such as rooflights, vents, etc.
2. Mark on the roof plan the areas where leaks are suspected, based on the evidence obtained from within the building.
3. Taking all necessary safety precautions as detailed above for roof access and work, locate these points or areas on the roof, taking care to use roof boards lest the foot traffic should cause further problems. Proceed to examine roof sheets, fixings, flashings, rooflights and any other features in the zones concerned.
4. The most common causes for leakage on new profiled roofs are included in the following check-list:
   (a) loose fasteners, due to threads stripping or missing the support spacer, etc.;
   (b) over-tightened fasteners or ones which have been inserted at an angle so that the washer is not sealed;
   (c) side lap fasteners (particularly rivets) which have expanded between the top and bottom sheets leaving an open, unsecured lap;
   (d) sealant incorrectly applied or positioned, or used in insufficient quantity; failure of seal due to use of incorrect material, or sealant omitted altogether; sealant too firm when used causing roof-sheet to bow between fasteners;
   (e) damage to the locking side lap on secret-fix metal roof sheets or distortion at the concealed fasteners;
   (f) profile closures omitted, not properly secured or damaged;
   (g) inadequate end laps or side laps not nesting correctly;
   (h) fasteners incorrectly spaced;
   (i) incorrect selection of fastener type, size or washer;
   (j) use of open-cored rivets or ones with broken mandrel;
   (k) cracked rooflight sheets due to thermal movement, overtightened fasteners or incorrect type of fastener;
   (l) incorrectly dressed or fractured flashings at ridge, verges, eaves, flues, vents and other penetrations of the roof surface;
   (m) roof sheets incorrectly laid, split or other wise defective.
5. Additional causes for leakage on older profiled roofs are included in the following check-list:
   (a) fasteners with oversized or elongated holes caused by incorrect drilling or thermal movement of the sheeting;
   (b) badly corroded fasteners;
   (c) where fixing into timber is used, the fasteners can jack-up or become loose due to shrinkage of the timber;
   (d) insulation quilts and foam boards laid over spacers can compact, loosening the fastener washer.

BRITISH AND EUROPEAN STANDARDS

BS 460: 1964   Specification for cast iron rainwater goods

BS 476: Part 3: 1958   External fire exposure roof test (withdrawn but included in BS 5427)

BS 476: Part 3: 1975   External fire exposure roof test

BS 476: Part 4   Non-combustibility test for materials

BS 476: Part 6   Method of test for fire propagation for products

BS 476: Part 7   Method of test to determine the classification of the surface spread of flame of products

BS 476: Part 11   Method for assessing the heat emission from building materials

BS 476: Part 20   Method for determination of the fire resistance of elements of construction (general principles)

BS 476: Part 22   Method for determination of the fire resistance of non-loadbearing elements of construction

BS 569: 1973   Specification for asbestos-cement rainwater goods

BS 729: 1971   Specification for hot dip galvanized coatings on iron and steel articles

BS 970: Part 1: 1996   General inspection and testing procedures and specific requirements for carbon, carbon manganese, alloy and stainless steels

BS 1091: 1963   Specification for pressed steel gutters, rainwater pipes, fittings and accessories (obsolescent)

BS 1178: 1982   Specification for milled lead sheet for building purposes

BS 1449: Part 2: 1983   Specification for stainless and heat-resisting steel plate, sheet and strip

BS 2782: 1970   Methods of testing plastics

BS 4154: Part 1: 1985   GRP – Specification for material and performance requirements

BS 4203: Part 2: 1980   PVC – Specification for profiles and dimensions

BS 4576: Part 1: 1989   PVC-U – Half-round gutters and pipes of circular cross-section

BS 5250: 1989   Code of Practice for control of condensation in buildings

BS 5427: Part 1: 1996   Code of Practice for the use of profiled sheet for roof and wall cladding on buildings

BS 5534: Part 1: 1997   Code of Practice for slating and tiling

BS 5588: Part II: 1997	Fire precautions in the design – Code of Practice for shops, offices, industrial, storage and other similar buildings
BS 6181: 1981	Method of test for air permeability of joints in building
BS 6367: 1983	Code of Practice for drainage of roofs and paved areas
BS 6399: Part 2: 1997	Code of Practice for wind loads
BS 6399: Part 3: 1988	Code of Practice for imposed roof loads
BS 7543: 1992	Guide to durability of buildings and building elements, products and components
BS 8217: 1994	Code of Practice for built-up felt roofing
BS 8200: 1985	Code of Practice for design of non-loadbearing external vertical enclosures of buildings
CP 3: Chapter V: Part 2: 1972	Wind loads (obsolescent)
CP 143: Part 5: 1964	Code of Practice – zinc
BS EN 494: 1994	Fibre-cement profiled sheets and fittings for roofing
prEN 508–1	Product standard for profiled steel sheeting
prEN 508–2	Product standard for profiled aluminium sheeting
prEN 508–3	Product standard for profiled stainless steel
BS EN 607: 1996	Eaves, gutters and fittings made of PVC-U
prEN 1187	Fire test – external fire exposure of roof
BS EN 1263–1: 1997	Safety nets: safety requirements test method
BS EN 1263–2: 1998	Safety nets: safety requirements for erection of safety nets
BS EN 1462: 1997	Brackets for eaves gutters. Requirements and testing
BS EN 1013–1: 1998	Light transmitting profiled plastic sheeting for single skin roofing. General requirements and test methods
BS EN 1013–2: 1999	Specific requirements and test methods for sheets of GRP
BS EN 1013–3: 1998	Specific requirements and test methods for PVC
prEN 1013–4	Specific requirements and test methods for PC
prEN 1013–5	Specific requirements and test methods for PMMA
BS EN 516: 1995	Prefabricated assessories for roofing. Installations for roof access. Walkways, treads and steps
BS EN 517: 1995	Prefabricated accessories for roofing. Road safety hooks
BS EN 612: 1996	Eaves gutters and rainwater down pipes of metal sheet. Definition, classification and requirements
BS EN ISO 9001	Quality systems. Model for quality assurance in design, development, production, installation and servicing
BS EN ISO 9002	Quality systems. Model for quality assurance in production, installation and servicing

The above British and European Standards may be purchased from:
British Standards Institution
389 Chiswick High Road
London, W4 4AL

B

REFERENCE DOCUMENTS:
The Building Regulations

Approved documents
B – fire safety
F – means of ventilation/condensation in roofs
L – conservation of fuel and power
These may be purchased from: HMSO Publications Centre, PO Box 276, London SW8 5DT.

MCRMA – Technical Papers
No.1   Daylighting recommended good practice in metal clad light industrial buildings
No. 2   Curved sheeting manual
No. 3   Secret fix roofing design guide
No. 4   Fire and external steel-clad walls: guidance notes to the revised Building Regulations, 1992
No. 5   Metal wall cladding detailing guide
No. 6   Profiled metal roofing design guide
No. 7   Fire design of steel sheet clad external walls for building; construction performance standards and design
No. 8   Acoustic design guide for metal roof and wall cladding

No. 9   Composite roof and wall cladding panel design guide
No.10   Profiled metal cladding for roofs and walls; guidance notes on revised Building Regulations 1995 – Parts L and F
No.11   Flashings for metal roof and wall cladding; design, detailing and installation guide

The above MCRMA technical papers may be purchased from:
The Metal Cladding & Roofing Manufacturers Association Ltd
18 Mere Farm Road, Noctorum
Birkenhead, Merseyside, L43 9TT

INDEX

Access and safety 98
Accommodating thermal movement 6
Air leakage testing 20
Air leakage 19
   metal cladding 24
Air, moisture and condensation 13
Appearance
   metal cladding 21
   fibre-cement 53
Application of sealant 86
Asbestos cement 53
Assessment of a systems thermal performance 17
   metal cladding 30
Assessment of the U–value for a metal cladding system 30
Avoiding bimetallic contact 8

Bimetallic corrosion 7, 8
Breather membrane 15, 33–4
British and European Standards 105
British Standards for fibre-cement products 54, 105
BS 6367: 1983 Code of Practice for drainage for roofs 88
Building Regulations 107
   Approved document B – fire 9
   Approved document F – ventilation 12
   Approved document L – air leakage 19
Built-up metal roofing systems 30
   built-up systems with spacers 30
   breather membranes 33–4
   curved sheeting 37
   horizontal cladding 39
   hybrid construction 34
   internal humidity grades 34
   reduced tolerance for fitting, curves 34
   secret-fix sheeting 35
   side and end laps 26, 35

   U-value 30
   vapour control measures 32

Calculation of flows in gutters 90
Calculation of U-values
   general 17
   profiled metal 30
   fibre-cement 64
CDM regulations 99
CEN fire regulations 10
Composite panels 42
   air leakage through the building envelope 19
   batch production 43
   concealed fix composite panels 46
   continuous production 43
   control of condensation 44
   eaves 48
   finishes for wall panel 49
   fire 45
   fixing requirements 51
   health and safety 51
   individual panel production 44
   inspection and maintenance 52
   insulation core 43
   loadings 45
   manufacturing 43
   metal facings 42
   openings 48
   panel performance 44
   purlin/rail support 46
   purpose made wall panels 49
   repairs to damaged panels 51
   ridges 48
   roof pitch and sheeting laps 46
   roof systems as wall cladding 49

roof applications 46
rooflights 49
side and end laps 46, 47
site storage 51
special details 50
structural 44, 46
thermal 44
traditional composite panels 46
U-value 44
verges 48
wall cladding 49
Commencement of work on site 102
Composite gutters 94
Concealed fix sheeting
  built-up metal 36
  composites 47
Condensation, the role of the designer and contractor
  12
Construction (Design and Management) Regulations
  1994 100
Construction (Health, Safety and Welfare) Regulations
  1996 99
Construction, safety and maintenance 97
Curved sheeting 37

Day lighting levels 67
Dead load 3, 31
Design considerations, general performance 3
Design of outlets, gutters 91
Design rates of rainfall 88
Designer's responsibilities
  performance under loads 3
  fasteners 81
  gutters 88
  safety 97
  vapour control 14
Deterioration, material incompatibility 7
Durability
  sheeting 21
  metal 22
  fibre-cement 64
  rooflights 67

Eaves gutters 93
Eaves closure foam fillers 87
Employees' duties, safety 98
Employer duties, safety 98
End laps
  metal sheeting 26, 35
  composite panels 46, 47
  fibre-cement 61
  rooflights 73
Existing roofs, safety 98
Exposed fasteners, colour matching 80
External gutters 93

Fastners and Fixings 77
  choice of materials 78
  designers' responsibilities 81
  fastener washers 80
  fastener heads 80
  fastener application 81
  fastener materials 78
  forces acting on fasteners 81
  modes of failure 75
  other fixing types 78

primary fasteners 75
secondary fasteners 75
sheeting contractor's responsibilities 81
Fault finding analysis 103
Fibre-cement sheeting 53
  appearance 53
  British Standards 54
  chemical resistance 64
  coloured fibre-cement and metal lining sheets 63
  condensation 65
  design guidance procedure 55
  design loading 56
  double skin system 64
  effect of frost 65
  fixing 57
  installation 57
  material characteristics 64
  metal lining system 64
  mitred corners, fixing procedure for laying with 61
  movement joints 55
  natural grey sheets 62
  overhangs 61
  preparatory safety 63
  properties 54
  recommendations for installation 55
  rigid installation boards 64
  roofing systems 64
  safety at work 63
  sound insulation 64
  thermal movement 64
  U-values 64
  weather protection 58
Final checking 103
Fire 9
  fire requirements, updating 11
  fire requirements for walls 11
  fire performance
    composites 45
    rooflights 11, 68
  fire resistant closures 87
Fittings, fibre-cement 58–60
FM approval 12
Functions of sealants 82

Gutters
  aluminium 94
  asbestos cement 95
  cast iron 95
  coatings for mild steel and aluminium 95
  fixing 92
  general principles of design 88
  glass fibre reinforced plastic 95
  insulation 94
  installation 96
  joints 93
  leaks 96
  maintenance of gutters 96
  materials for gutters and rainwater pipes 94
  plastisol coated steel 95
  shapes 92
  stainless steel 95
  stains 96
  steel 94
  syphonic outlet systems 91
  thermal requirements 94
  valley, parapet and boundary wall 92

workmanship 96
  see also Rainwater goods 88

Handling 101
Health and safety 99
Health and Safety at Work, etc., Act 1974 97, 99
Horizontally laid cladding 39
Humidity grades 24
Hybrid construction 24

Imposed load 3
Insulated composite panels 42
Insulated fibre-cement 63
Insulated gutters 94
Insulated construction *see* chapter on sheeting type
Internal environment 12, 47

Lap treatments
  see end laps
  see side laps
Latent deterioration 9
Leaks 96
Life to first maintenance 22
Life expectancy of profiled metal 22
Light weight profiled cladding 1
  common requirements for 1
  condensation, role of designer and contractor 12
  loads on lining sheets 3
  performance of roofing and cladding 4
  performance under load 4
  performance of profiled sheeting 3
  surface protection 8
  temperature range, table of typical ranges 6
  thermal movement 5
Loss Prevention Council (LPC) 12

Maintenance inspections 102
Maintenance 102
  see also chapter on sheeting type
Maintenance manual 102
Management of Health and Safety at Work Regulations 1992 99
Method of test for air permeability of joints in building 20
Methods of joining
  gutters 93
Movement joints
  gutters 93
  fibre-cement sheets 55
Movement of moisture through construction 13

Notes for guidance on storage 101
  see also chapter on sheeting type

Openings in roofing/vertical 4

Performance under load 4
Performance of profiled sheeting 3
Performance of roofing and cladding 4
Profiled metal roof and wall cladding 21
  aluminium substrate 22
  applications 23
  choice of profile 23
  condensation control 24
  design of flashings 27, 28
  durability 22
  functional life 23

hybrid construction 24
  internal humidity grades 24
  period to repaint decision 22
  selection by the specifier 23
  standards for profiled sheeting 22
  statutory and other requirements 21
  steel substrate 22
Plant and scaffold registers 98
Preparation 102
Products included in the guide 1
Profile closures 60, 86
Profile fillers 86
Profiled sheet metal 21
Purlin/side rail restraint and thickness 5, 46

Rooflights 66
  building regulations 68
  design aspects 66
  durability 67
  fasteners 71, 74
  loading, general 69
  material and types 66
  maintenance 74
  safety and CDM 70
  sealants 73
  thermal expansion 74
  transport, handling and storage 74
  typical application 67
Rainwater goods 88
  cast iron 95
  general principles of design 88
  gutters aluminium 94
  gutters coatings for mild steel and aluminium 95
  gutter fixing 92
  gutters glass fibre reinforced plastic 95
  gutter insulation 94
  gutter joint 93
  gutters plastisol coated steel 95
  gutter shapes 92
  gutters stainless 95
  gutters steel 94
  gutter thermal requirements 94
  gutters valley, parapet and boundary wall 92
  maintenance of gutters 96
  materials for gutters and rainwater pipes 94
Recommendations for the design of metal flashings 28
Reducing the risk of condensation 12
Reference Documents 107
Repairs to damaged composite panels 51
Roof penetrations 28

Safe systems of work, fibre-cement 63
Safety and CDM 99
Sealants and profile fillers 82
  functions and properties required of sealants 84
  reasons for use of sealants 82
Secret-fix roofing systems 35
Sheeting contractor's responsibilities
  choice of cladding 100
  condensation 12
  fasteners 81
  site Safety 97
Side laps
  metal sheeting 26, 35
  composite panels 47

fibre-cement 61
   rooflights 73
Site safety 97
   see also chapter on sheeting type
Site handling 101
   see also chapter on sheeting type
Site practice and installing 100
   see also chapter on sheeting type
Site safety meetings 98
Snow Loading 3
Spacers, spacer bars 30, 31
Staining 96
Standards for profiled sheeting 22
Statutory and other requirements 21
Steel substrate 22
Storage 101
   see also chapter on sheeting type
Stressed skin performance 5
Sub-framing 39
Surface protection 8
Syphonic outlet systems 91

Thermal movement 5
Temperature range, table of typical ranges 6
Thermal insulation U-values 68
Thermal insulation, physical properties 16, 43

Thermal expansion 5
   see also chapter on sheeting type
Training 98
Transport, handling and storage 100
   see also chapter on sheeting type
Types of metal substrate 22

Updating fire requirements 11
Use of breather membrane 15
Use of light colours 23
Use of a VCL 14
U-value 17, 30, 44, 64, 68

Vapour control and acoustic absorption 35
Vapour control and rooflights 69
Vapour Control Layers (VCL) 14
Vapour control with insulated built-up metal
   sheeting 32–3
Vented fillers 87
Ventilation - Approved Document F 12

Welfare 99
Wind forces 3
Wind load, responsibility for load calculation 3

Zed spacers 31